BATTLE RHYTHM

THE MILITARY INSPIRED PERSONAL PLANNING, DISCIPLINE AND MOTIVATION GUIDE

JOSH FRANCIS

Battle Rhythm: The military inspired personal planning, discipline and motivation guide © 2019
Josh Francis (Red Diamond Consultancy)

ISBN: 978-1-922264-77-0 (paperback)

Cataloguing-in-Publication information for this title is listed with the National Library of Australia.

Published in Australia by **Josh Francis (Red Diamond Consultancy)** and InHouse Publishing.
www.red-diamond.com.au

Printed in Australia by InHouse Print & Design.
www.inhousepublishing.com.au

Contents

Introduction

"If you find yourself in a fair fight, you didn't plan your mission properly"

DAVID HACKWORTH

The modern world is full of numerous distractions that can hinder us from being able to achieve what we set out to do. The rapid rise in technology was designed to make our lives more efficient. The development of the internet and smartphones has allowed us to live our lives at the touch of a button. This technology means that we have access to more things that can occupy (or waste) our time, such as ordering food, watching television and taking out health care, which can now all be done at the touch of a button. However, this has meant that the pace with which society operates has risen exponentially, as people try to keep up by doing more without more time in which to do it.

This can lead to a lot of wasted effort as people struggle to use their limited time correctly, and without distraction, when setting out to do the things they want to do. As a consequence, it is possible (as happens frequently) that people spend quite a lot of time doing something without actually achieving anything. This is due to two reasons—firstly, a lack of proper planning and time management skills and secondly, a lack of discipline and motivation. These two things are inextricably linked. Someone can have great personal planning skills, but it's pointless having a good plan if it's not followed, not made adaptable to changing circumstances or if the time available is used inappropriately. Similarly, the most disciplined and motivated person can end up achieving nothing if they waste their time due to poor personal planning and time management.

The Australian and US military forces are renowned for their meticulous organisational skills, honed by decades of war and peace during the 20th and 21st centuries. During this time, literally thousands of soldiers, sailors, marines and airmen and airwomen have been, and continue to be, sent across the world, with all the logistical requirements, to conduct military operations. To be able to do this requires considerable planning and time management skills, as well as significant discipline and motivation from those undertaking these tasks. Particularly during the past fifteen years, these militaries have been constantly deployed to numerous global conflicts stemming from the 2001 attacks on the World Trade Centre in New York City, as well as to peacekeeping and humanitarian-assistance efforts. This has been enabled by the high-quality level of training that is provided to the resilient, highly disciplined and motivated service personnel who employ knowledge, skills and attitudes (KSAs) instilled in them through military service, often in very difficult circumstances.

At a cumulative level, these KSAs allow the military to perform highly demanding tasks under arduous conditions, ensuring mission success— measured as the achievement of all set military objectives. In recent times, tens of thousands of servicemen and women from numerous countries have left their respective branch of the military to pursue a life in what is commonly called 'civvy street'—the civilian workforce. More and more, their KSAs are being recognised and utilised by everyone, from small businesses to large corporations, who see the beneficial applications that those who have undertaken military service can provide in the workplace.

From the moment a prospective soldier enters their recruit training establishment (in US parlance 'bootcamp') they are taught the very basics of discipline and time management. The arduous hours spent making beds, polishing boots, marching up and down the parade ground or sitting in a trench half-full of water in the middle of a cold night all impart KSAs that may one day be called upon in conflict.

The military, in both war and peacetime, operates in a completely different environment from the 'normal' world. In private business, for example, a lack of planning, ill-discipline and poor time management can result in a loss of client business, a reduction in profits or reputational harm. All negative consequences in the context of the environment the business operates in. On a personal level, a failure to manage your own time or plans can mean missing the bus to work or not having your taxes prepared and ready to submit at the end of the financial year, which may result in outcomes with varying levels of consequence. A business might even shut down as a result; however, it is not expected that injuries to people will occur. In the military, a failure to conduct meticulous planning, remain disciplined and adhere to designated timings can result in injuries or deaths. This is the extreme but realistic nature of all military operations, and as such, those who operate within it quickly become experts in matters of planning and discipline.

This handbook is a guide, designed to show you how to apply some of those military-style skills and methods in your own personal planning and time management. Additionally, it will show you how to instil self-discipline in yourself, and how to leverage your motivations to remain disciplined when it feels like the going is getting tough, so that you can stay committed to the tasks you have set out to undertake. This will allow you to plan correctly, be more efficient in your time management, and ultimately become disciplined and motivated enough to achieve your goals.

Firstly, we will look at the '**C2 Framework**'. This is a personal planning and time management methodology. It will require you to critically review and break down your current commitments and use of time, with the aim of allocating it more efficiently in order to achieve your goals. Hopefully, you'll be surprised (or perhaps shocked) at how much time opens up to you (and how much you have been wasting) when you look at it closely and seek to use it better. It will show you some basic planning considerations, similar to those used by numerous militaries worldwide. Using these planning tools will result in a more productive self, without the requirement to have a drill sergeant yelling at you to get things done!

The second part will look at discipline and motivation. Great planning is for naught if you aren't resilient when the plan needs, or is forced, to change.

Everyone is motivated by different things. The key is to remain disciplined enough to keep going when that motivation is either wavering or disappearing altogether. Applying some of the basic tenets of military discipline in your own life—leveraging off your hard and soft motivations—will help you achieve the things you never thought possible and make everyday living more organised and less stressful.

Don't be afraid to give this methodology a go. These skills utilised by the military have been polished over several decades of operations globally. It won't turn you into a rigid soldier who forgets how to enjoy life. Quite the opposite. It will act as a baseline which will allow you to confidently plan for the things you want to achieve. As the framework is flexible, it will allow you to make changes without having to recreate the whole plan. Indeed, the modern soldier is renowned for their independence, their ability to think creatively, to show leadership under pressure, and to thrive as an individual, or as part of a team, in environments that would make most people recoil in fear. This handbook will allow you to use these hard-learnt lessons so that you can become more productive in your own life and can achieve your own mission success!

PART 1
THE C2 FRAMEWORK

The C2 Framework

"There are no secrets to success.
It is the result of preparation, hard work, and learning from failure"

COLIN POWELL

E veryone has goals that they want to achieve and things that they want to do, but one of the most common excuses as to why people don't achieve them is 'I just don't have enough time'. You'd be surprised at just how many people claim they don't have enough time to do everything they want in life. But, when you examine closely how they've used their time, you quickly see that, with a little forethought, guidance and better planning, it would have been easy to fit in that extra gym session, to have met that work deadline or to have attended that school concert. This is what is called 'effective time management'.

In military terminology, C2 is the abbreviation for command and control. Every military unit, no matter its size, has a C2 function with a designated person who is given the sole authority over the people and resources underneath them. They are also responsible for planning and conducting operations. The '**C2 Framework**' is a planning methodology developed to allow individuals to gain more control over their own lives. By using the principles commonly used in military planning, they can better plan for the things they want to do and the goals they want to achieve. It is an easy-to-manage tool that will give you the structure and the guidance you need to achieve your goals and manage your time more efficiently. It will reduce stress as proper personal planning gives guidance that removes ambiguity and confusion in your own life. It will show you how to use your time wisely. Figure 1 outlines what it looks like:

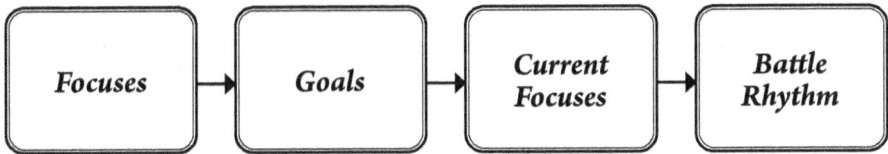

| Focuses | → | Goals | → | Current Focuses | → | Battle Rhythm |

Figure 1—The C2 Framework

The C2 Framework is broken down into four parts:
- Focuses—these are the division of all the commitments that take up the time in your life;
- Goals—all the things you want to do and achieve (divided among these focuses);
- Current Focus—the immediate tasks you are working on based on your goals; and finally, but most importantly,
- Battle Rhythm—the personal routine where you actually assign the time to do the things you need and want to do to achieve these goals.

Every part of the C2 Framework leads into the next part (i.e. your focuses will influence your goals, which will be prioritised in current focus, which will be worked towards at a time determined by your battle rhythm). Because the battle rhythm is vital to efficiently allocating time to working towards your goals, it will be explained in detail first, as it will determine just how much time you have available, once all your current and future commitments are considered, to work towards your goals.

We will analyse these four parts individually, examine how they are similar to military planning, and look at time management and why, as a combined framework, it is effective. From here, we will look at how you can use this methodology to start establishing your own framework to suit your own ambitions. Once you have set-up your own framework, it won't need to be managed too often, as you will assign specific time to review it and track your progress. All you will need to do is simply add new tasks, remove others that have either been completed or are no longer viable, and make changes to your focuses, goals and routines depending on your changing life circumstances.

A strategy is a plan of action designed to achieve a long term or overall aim. Whether it be for the conduct of complex warfighting or for more routine activities such as equipment procurement, the military will develop a strategy to guide for the planning and execution of the respective task. A strategy can take some detailed initial planning, but once complete, it often requires minimal management and amendments in order to maintain it and track its progress. The C2 Framework will be your methodology. It will allow you to plan a personal life strategy so you can set out to achieve goals that might be related or totally different from each other. The strategy will involve identifying where your goals sit within your focuses, prioritising when they need to be done, determining how they can be achieved, then setting out to undertake the tasks necessary to achieve them in a considered and orderly manner by using your time in the most efficient way possible. By adhering to this methodology, you will plan your life more effectively, begin to use your time more efficiently, and will ultimately be in a better position to achieve your goals and ambitions.

SUMMARY POINTS
- **The C2 Framework is a military-inspired methodology aimed to help you plan your life and manage your time more efficiently.**
- **It consists of: Focuses → Goals → Current Focuses → Battle Rhythm**
- **It will help you develop and plan your strategy for life.**

Battle Rhythm

"The only easy day was yesterday"

US NAVY SEALS

The hard fact is that there are only twenty-four hours in a day, seven days in a week, four (thereabouts) weeks in a month, and fifty-two weeks in a year. Depending on your own personal circumstances, time can feel like it goes very fast or very slow. Most people would agree with the old saying, 'time flies when you're having fun'. The opposite is also true. Ten minutes at a party will go by very quickly if it's a great party. However, try doing push-ups continually for ten minutes … that makes time feel very slow. So, while time is relative, it's also limited. It comes and goes, whether you are ready for it or not.

At the lowest level of the C2 Framework is the '**Battle Rhythm**'. A battle rhythm is also known as a routine. This is an invaluable time management tool that must be established first. Your own battle rhythm will dictate how much time is available to you to undertake the tasks you set out to accomplish and will, therefore, allow you to plan more efficiently. It's pointless allocating only two hours to do something that experience suggests takes four hours, just as it's inefficient to allocate half a day to an activity that only takes an hour or two. In military circles, a battle rhythm is the allotment and prioritisation of time to conduct the tasks required for the efficient running of military operations, whether it be at a small team level, right up to a deployed battle-group consisting of hundreds and thousands of soldiers. It is essentially the daily routine (key point being that it's set up to run daily) where timings for the conduct of different things are laid out, usually to the minute, to ensure operational requirements are met. Figure 2 outlines the 'Battle Rhythm' within the C2 Framework.

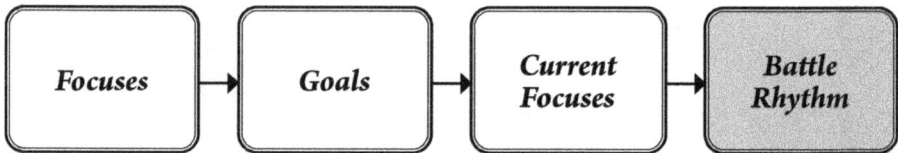

```
┌──────────┐    ┌──────────┐    ┌──────────┐    ┌──────────┐
│ Focuses  │──▶ │  Goals   │──▶ │ Current  │──▶ │  Battle  │
│          │    │          │    │ Focuses  │    │  Rhythm  │
└──────────┘    └──────────┘    └──────────┘    └──────────┘
```

Figure 2—Battle Rhythm with C2

The battle rhythm determines when and what you do with your time. The remainder of the C2 Framework is all about the planning; whereas, the battle rhythm is the execution of that plan. This is why I am discussing it first so, as you read about the other parts of the methodology, you are consciously considering the development of your own framework. All planning is undertaken with an appreciation of the time that will be available to undertake that plan, so your own planning will be based on your own routines (which can be developed or adapted as needed, based on what it is you want to do and achieve).

Time is a commodity and should be treated like a valuable resource. You can't make more of it, you can only use what time you have available more efficiently. You should treat it the same way you would money, ensuring that you undertake a careful cost/benefit analysis before expending it. To enable you to manage your time correctly and be successful in achieving your goals, it is critical that you develop your own battle rhythm. This is where you actually designate the time in which you will do the things you need to do, so that you don't spend days watching time tick by without actually achieving anything.

However, to create a good battle rhythm, that is both efficient and effective, you must consider both the things you *need* to do and *want* to do. When possible, a military unit will run a battle rhythm that suits their specific requirements. During the 1982 Falklands War, the British Task Force, under Rear Admiral Sandy Woodward, ran a battle rhythm that was in line with the time zone back in the United Kingdom. As such, the task force's day began in the dark early hours of the morning, and this allowed for their communication to be synchronised with the working hours of the Ministry of Defence at Whitehall in London. That specific battle rhythm suited their needs at the time, even though it was different to what they would have been used to when operating in peacetime. Various military units run set battle rhythms that don't change regardless of the situation. Whether on operations or during peacetime, Australian submarines run what's known as a 'six on, six off' routine whilst at sea. All personnel will spend six hours on watch, performing their trade whether it be on the bridge, in the operations centre or in the engine room. They will then have six hours off, during which time they perform administrative tasks, eat and try to find some sleep. This routine repeats itself twenty-four hours a day, seven days a week while they are at sea.

However, there will be factors which will influence your battle rhythm, some of which you will have no control over. Your employment, for example, may not permit you to tell the boss that you would rather finish at 3 p.m. to play golf! It has designated hours that, withstanding individual flexible arrangements, may not be able to be moved around. During the Australian Army's operation to train Iraqi forces to fight ISIS, the Training Team Commander had to set a battle rhythm that was in line with the cultural and military practices of the Iraqi soldiers his unit was training. This meant that training occurred from well before dawn up until midday, to avoid the stifling heat of the latter half of the day. Additionally, no training occurred on Fridays, in line with the Arabic weekend. This was not how Western armies usually operated, but it was required for that particular mission.

When establishing your own daily battle rhythm, look at how you are currently using your time. Be honest, be critical and be realistic. It will be a daily battle rhythm, though you can look to produce a weekly one later on. Include even the most basic things, such as how much time you spend looking at emails at work, doing online shopping, sitting outside having coffees and talking to friends or even sitting on the bench press while at the gym not actually pressing anything! Start from the time you get up in the morning, all the way through to when you go to bed. Don't forget little things like when you have breakfast, as this will ensure time is allocated for everything you need to do (we all need to eat). Look at all the tasks you are currently doing, and how long you are spending on them. Also, look at what you do while at work, as this will likely take up the majority of your time each day and will likely have a battle rhythm in itself that you follow.

Write it down so you can see on paper where your time is actually being spent. Be specific. Write down how long each activity currently takes (not how long it *should* take, that comes next), and when you start the next one. Now you can carefully consider how long it should take you. You are likely to be surprised, or shocked, at how much time you have been wasting, and how much time you could actually have available each day that you could be using more efficiently. Accountants get their clients to do the same with financial spending, and it's only when they see the actual numbers on paper that people realise just how much money they are wasting. The same applies to time.

Even by just writing out a basic outline, you can see a battle rhythm developing in front of you. This is far easier to manage than not allocating times for your daily tasks, trying to remember everything you were supposed to do, and then running out of time anyway! From here, you should seek to identify where you have significant gaps that could be utilised to do things or where you could move tasks around to avoid conflicts in your time. A battle rhythm works best when broken down into distinct blocks of time. This is how most corporations plan their business day, as it forces them to adhere to timeframes which prevent meetings from consistently going overtime. Billionaire entrepreneur Elon Musk is a fan of the 'pomodoro technique', whereby his day is broken down into blocks of twenty-five minutes each. The US president, with such a demanding schedule, typically breaks his schedule into fifteen-minute blocks, managed by a personal secretary who ensures that his time is used efficiently and effectively. Some of your tasks will use several blocks of time. Most of us don't have personal secretaries or the need to plan our days down to the minute, but using general blocks of time will guide your tasks each day and will help to confine them to set timeframes as opposed to having numerous open-ended tasks which can quickly turn what should be an eight-hour day into a fifteen-hour day.

Even if you do shiftwork, or are lucky enough to do freelance work where you can set your own hours, a battle rhythm can be set up for the hours you are awake, regardless of what time of the day or night that is. A battle rhythm can be a twenty-four-hour cycle, so it would include when you are asleep. Your battle rhythm may be different on weekends than it is on weekdays. Even on weekdays, there will be some days where you do certain things and not others, based on events that only occur weekly or monthly, such as tennis practice or after-work drinks. I recommend a daily, as well as a weekly, battle rhythm, as this will allow you to allot time to cover all the working days as well as the weekend, and it can account for these variations in the use of your time. However, establish one to suit your own requirements that will best allow you to use your time most efficiently. When planning what you want to do each day and week, the time available in which to do it will have been established.

There are two principles you must consider when setting your own battle rhythm. Is it *practical?* And, is it *flexible?* Be critical when looking at your time available, to ensure that it is actually feasible to do the things you want to do within the allocated timeframe. You may have allocated two hours to spend at the gym, but are you really using that two hours? Could you achieve the same in one hour, thus freeing up that second hour to do something else? Similarly, if you are only allocating thirty minutes to go to the shops, but you know it takes at least an hour once you include all the traffic hassles, then you will end up eating into the time allocated to the next tasks in your battle rhythm. Consequently, you may not be able to catch up without depriving yourself of that good night's sleep or missing out on doing something else. This isn't very effective time management.

As a result, you may need to get up an extra half-hour earlier to fit in all the things you want achieve. This is called making time. Look at where you can multitask. For example, add a trip to the shops on the end of the gym session, so you don't have to head out again later. Some dry cleaners will do same-day service, and if you ask them, they may be able to complete the cleaning in an hour or two, so you might be able to do something while you are near the cleaners, saving time later by not having to return to the shop to collect your goods.

Be considerate about where you place the tasks within your battle rhythm that you do have control over. Knowing your own strengths and weaknesses is vital to creating a battle rhythm that is conducive to making the most efficient use of your time. Do you really want to leave studying for your exam until late at night when you are completely disinterested and too tired? Will you actually get up at 5 a.m. to do that gym session? Sometimes this may be the only time available to do those things, but at least you have given it some thought. Many successful business leaders talk about how they have a daily routine which exploits when they are feeling motivated and creative, yet still utilises the time efficiently even when they are tired. Many successful tech startups plan their meetings for in the mornings when people are fresh and at their most creative, whilst they leave the mundane, procedural-type activities for the afternoon when tiredness stifles creative thinking.

Attempt to leave some time available each day as spare time. You must retain the ability to move tasks around as required, because things will always come up at short notice, so your battle rhythm needs to maintain an element of flexibility where possible. You can achieve this by not unnecessarily cramming too much into your day and by deliberately allotting some time to yourself for doing things of your own choosing. If you then choose to use this spare time to work on a task, that's great, but it is mainly to be used to allow your battle rhythm to be flexible. You may choose to use this time to simply relax and take a breather. Rest is a vital part of any routine, as working too hard for too long will result in mental and physical fatigue which will only be detrimental to your own well-being, and it will hurt your ability to follow your plan.

Why do you need to make spare time? Consider the following: What happens if you, or a loved one, gets sick and you have to suddenly drop everything to deal with the situation? You will be grateful that you have some spare time available to catch up on your tasks. Similarly, you may be working on something that you are really getting drawn into, such as a study period, and this will allow you to go over the time you allocated for it if needed. Also, everyone needs time to just unwind and clear their head, otherwise you'll find yourself burning out very quickly. It might be some allocated time each day, or each week, but I highly encourage you to place some time in there for just you!

Leaving some free time in your battle rhythm will ensure there will be gaps which will allow you to adjust, or add, tasks if required. If you are a particularly busy person, you will be more stringent about how you allot your time. You may realise that there are simply not enough hours in a day (something I'm sure we have all said at some point), and will, therefore, have to examine ways to make time by using what is available more efficiently. As we look at planning later in the book, the blocks of time you have created will start to naturally fill themselves because there are timings which you have no control over, such as when you have to go to work, and subsequently, you will learn to prioritise your tasks.

Your battle rhythm will become vital to your personal planning and time management, because you will consider the things in your life you want to do and achieve and will use the battle rhythm to allocate the time to actually do it. A failure to adhere to it, or to adapt it when required, will result in mission failure—you will waste your time and not achieve your goals.

You may have to play around with it initially, until you find something that works for you, and what sort of things you want to put in it. You may work better with doing the same things at the same time each day and week, because this might be how you remain disciplined enough to get everything done. However, if you have the ability, consider mixing it up from time to time, perhaps by doing a gym session in the morning instead of the evening or going for a walk in the afternoon after work with your family. This can help prevent the feeling that your routine is becoming too mundane. Don't forget to remember to make some time for the miscellaneous tasks that are also part of life, such as watering the plants and paying the bills.

Record your battle rhythm any way you see fit. Some people write it in the front of their diary, others in their electronic calendar, whilst some people are simply able to remember it. The busier you become, the more important tracking your battle rhythm and allotting the times to the tasks will become, because you won't have time to spend your days figuring out what it is you are supposed to be doing hour-by-hour (especially if you don't have a personal secretary). If writing it down on the mirror in your bathroom works, then do that! Set your battle rhythm to work best for you, your own requirements and your personal desires, but always adhere to the principles. Looking at the rest of the C2 Framework will show why the battle rhythm is so important. A battle rhythm has been proven to work, specifically in high-pressure environments, because it provides direction for use of time. It is easily applicable to your own life, whether you are a very busy person or you have all the time in the world on your hands.

SUMMARY POINTS
- **Efficient time management is achieved through the establishment of a personal battle rhythm.**
- **Your battle rhythm must consider your own needs and requirements, as well as the timings of things you don't have control over (i.e. your work schedule).**
- **It must be *practical* and *flexible.***

WRITE DOWN AN OUTLINE FOR YOUR BATTLE RHYTHM

Focuses

"Soldiers can sometimes make decisions that are smarter than the orders they've been given"

ORSON SCOTT CARD

The second part of the C2 Framework will look at what I've labelled as 'Focuses'. These are the components in your life that you spend your time on, broken down into different areas (or themes). Focuses can be what you spend lots of time on or a small amount of time on, but regardless, they are still important and relevant in your life. Essentially, they are the things that make up your lifestyle. It will likely be obvious what they are, as they are the things that are most important to you. However, for the sake of better organisation, and thus better planning, you divide them up so they can be considered separately when looking at setting your goals and using your time. You can have as many or as few as you like, but just consider that the less you have, the easier it will be to manage them. They will be the headings under which your goals (and, subsequently, time) will be allocated. Having focuses ensures that you avoid the feeling of trying to do everything at once. It makes looking at how you use your time holistically less daunting, because it allows you to *focus* on one thing at a time. Figure 3 indicates 'Focuses' in the C2 Framework:

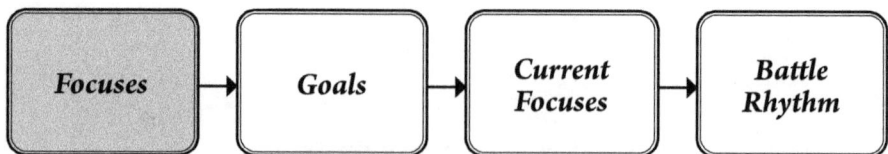

| Focuses | → | Goals | → | Current Focuses | → | Battle Rhythm |

Figure 3—Focuses with C2

Examples of what your focuses might be are numerous and will likely include a mix of things you both need to have and want to have. For the majority of adults, one of the foremost ones will be your employment (your job). For students, it will be school or university. These will likely take up quite a bit of your time, and how much time (and when) you devote to this focus will be determined by your workplace, school or university. You will have goals within this focus, be it training for higher qualifications or getting promoted. A second, and highly recommended, focus will be finance. Although linked to the employment focus, as you work to make money, it should still be a separate one as money management should always be given careful attention. Good financial management will involve setting goals and making the time to manage it correctly will ensure you don't waste this hard-earned resource. Finance management won't require much of your time, but inserting it into your planning will guarantee that you have devoted some thought to managing your finances correctly, and this will allow you to set goals associated with it (e.g. saving for a house deposit or a holiday). There are numerous resources available to assist with good personal finance management, so I will only pass one bit of advice that I received a long time ago: 'If you don't have it, don't spend it!'

Consider focuses that might not already be part of your life but that you have previously given some thought to. If you decide you want to become really fit, this will take time, and so a focus on fitness should be inserted into your planning. You will set goals and allocate time to this focus, just like any of the other ones.

Your personal interests, which you use some of your time on, could be bundled under one focus, which could be loosely titled 'hobbies'. One example could be that you want to devote more time to your family. If you are leading a particularly busy life, or are just worried that you are not spending enough time with your family in general, then make this a focus. Your subsequent goals may simply be spending more time doing things with your spouse, children or even pets! You need to consider how to make the time available to actively do this, so making it a focus means you consider this when conducting your personal planning. It's how you find time to fit a round of golf in or to do your Christmas shopping.

The specifics of setting goals will be described in the next chapter. The focuses are what you spend your time on, and they may change as your life circumstances change. They can last for six months or a year, or your focuses may be enduring (such as your employment focus). If you choose to, you could use the concept of 'Focuses' to have 'focuses within focuses'. For example, you could divide your workday into focuses, such as emails, personal management, professional development etc. If looking to use weights in your fitness focus, then your smaller focuses will be arms, legs, back, shoulders, cardio etc. You would then have goals within each of these smaller focuses.

Within an army, all the capabilities are divided into different components based on what each soldier's role is and the resources they require. All these capabilities are what make up the army. These are known as battlefield operating systems (BOS) concepts. BOS include infantry, air power and logistics, to name just a few. When conducting planning for an operation of any nature, ranging from peacekeeping through to armed conflict, an army will consider these BOS concepts as the key components that will need to be considered for inclusion when conducting the planning. From here, each component will have tasks assigned to it as part of any overall plan.

As the BOS make up an army, the focuses are what make up your life. Therefore, when doing your own personal planning, look at all the components in your life that you devote your time to. These are your BOS concepts. These are your focuses. When determining your goals and assigning your time, ensure that you have first considered each of your focuses individually to ensure that you haven't forgotten about any of them. When needed, you may add or remove a focus to your C2 Framework based on your own life choices or changing life circumstances. The military often adds or removes capabilities, depending on their own needs. The Royal New Zealand Air Force removed its air combat (jet fighter) capability from service in 2001, citing it as no longer relevant to the long-term development of the New Zealand Defence Force. Australia added an amphibious capability to the Royal Australian Navy (RAN) in 2014 with the commissioning of the helicopter landing ship HMAS *Canberra*, the largest-ever ship to serve in the RAN. You will inevitably do similar as your own life situation changes. For example, once you have completed a course of study, this will no longer be one of your focuses. It's likely that your new focus will be the full-time job that you just started, and with it will be the associated goals. Turning forty is usually the impetus for some people to start living a healthier lifestyle, so fitness may become a focus if it hasn't been before.

Once established, your focuses will simply reflect all the areas in your life you are currently devoting your time to, broken into easily defined (and managed) components. By having defined them, you can start looking at your goals in life, and you will find that they will fall under one of the focuses. Dividing your personal planning by focuses will start to make your life appear less cluttered, and it will allow you to start looking at your goals, and how you are going to achieve them, in a more structured manner.

SUMMARY POINTS

- **Focuses are the things you devote your time to within your life.**
- **Break them down so they are easily articulated based on what you need and want to devote your time to.**
- **They may be short term, long term or enduring. Add or remove them as best suits your needs.**

WRITE DOWN YOUR FOCUSES

Goals

"The journey of a thousand miles begins with a single step"

Lao Tzu

Your goals are things you want to achieve in life. Everyone has them. They are what drive our decisions and actions in life. When we are young, they are ambitious. Kids often dream of being astronauts, policemen, nurses or even to be a king or queen. Some of these goals get achieved, some do not. Not achieving our goals can be for a number of reasons; maybe because we didn't do well enough at school, or we find out we don't like cleaning up after sick people or because it turns out we aren't actually members of the Royal Family! Figure 4 outlines 'Goals' with the C2 Framework.

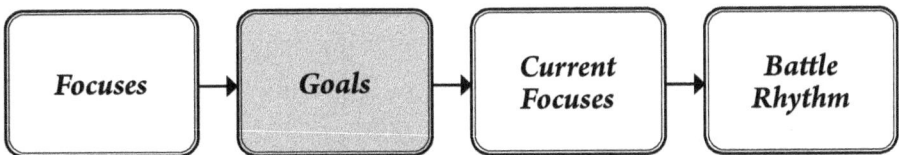

Focuses	→	Goals	→	Current Focuses	→	Battle Rhythm

Figure 4—Goals within C2

In military planning, a goal is known as an 'objective'—a single event or series of events that need to be met in order to achieve the plan. When making large scale plans, a military commander will have numerous objectives which need to be met in order to achieve a much larger one. These will be broken down into stages (often called 'phases'), so that it can be determined what needs to be done to complete the larger goal. Often, certain objectives need to be met before the next ones can be worked towards. This is known as the prerequisites. This is all established during the planning stages. When planning for the 1944 D-Day invasions in Normandy, France, the Allied commanders literally had thousands of smaller objectives, all of which needed to be achieved as a prerequisite before pursuing the next objective. In the early phases of the invasion, the British and American paratroopers were dropped in the middle of the night behind enemy lines to disrupt the German rear defences inland and prevent reinforcements from being able to move to the beachhead during the landings. Then, before dawn, the naval guns off the coast repeatedly shelled the beach defences to allow for the small craft, packed with infantrymen, to approach and land on the beaches. Once this occurred, the swarms of infantrymen spilled out from these crafts onto the sand to take control of the coastal defences. The result was a successful landing, and it was the beginning of the end of the Nazi's reign in Europe.

In normal terminology, smaller objectives are typically called tasks. Tasks combined together help you achieve larger goals (e.g. build a house), while some small goals may only have one main task (e.g. paint the bedroom; however, there will still be smaller tasks even within this small goal, such as buying the paint, preparing an undercoat, laying protective sheeting etc.).

Regardless of what they are, or how small or large they are, setting goals is critical in achieving things in your own life. It's the first step in establishing the direction you want to go, so that you can plan how to achieve your goals. When setting and planning your own goals, they will inevitably be based on one of your focuses as discussed in the previous chapter, so each goal will fall under one of your focuses. Without a plan, a goal is simply a wish!

When setting and planning goals, you must ask yourself five very basic questions:

- What do I want to achieve?
- Why do I want to achieve it?
- What do I need to do to achieve it?

- When am I going to achieve it? and;
- What will it look like at the end? (How do you measure success?)

In military planning, this is known as 'commander's intent'. This is where the commander of a military unit has:

- Examined what objectives they need to achieve as part of a wider plan (what they want to achieve);
- Looked at the purpose of achieving the goal (why it needs to be achieved);
- Determined the method in which they will do it (what needs to be done to achieve it);
- Decided when it be will undertaken (when they will do the various tasks to achieve it); and,
- Outlined what the final outcome should look like, known as an end state (measuring whether the operation was successful or not).

When looking at these five questions, the key one you need to look at to help in planning is what you need to do to achieve them. These are called 'decisive events'. A 'decisive event' is characterised as something which needs to be achieved as a prerequisite for achieving a military objective. When looking at your own goals, you have to determine what needs to be accomplished in order to achieve them. These are what you call 'tasks'. Outlining the tasks will allow you to prioritise them so that you can tangibly allocate the time you have available in order to undertake them.

When setting and planning your goals using the C2 Framework, you need to follow six essential principles. It is critical that you follow these principles; otherwise, your goals will lack direction. Known as SMARTA, they must be:

- Specific
- Measurable
- Attainable
- Relevant

- Time-bound
- Adaptable

Your goals need to be *specific* and related to your focus. A military commander goes into a campaign with a broad goal. It might simply be to win a war. However, smaller more specific goals need to be articulated within this in order to have the cumulative outcome that will see the achievement of the broader goal. Otherwise, it won't be clear on how it needs to be achieved. For example, one of your goals might be to achieve a healthier lifestyle. This is a common goal that most people set at some point in their life, often not long after midnight on the first of January! However, there are smaller goals (the tasks) you must undertake to achieve this. You must first establish what these tasks are, such as joining the gym, eating healthier, going for a walk after work, reducing your alcohol intake etc. All these specific tasks need to be planned, prioritised and achieved so your wider goal of a healthier lifestyle can be realised. A broad personal goal might be to live healthily and be happy in your relationships. The specific tasks will be to do physical fitness and to spend time with family and friends to help maintain the relationships you have established. This still requires time to do these things. This is an enduring goal, but it is specific in that there are particular things you need to keep doing in order to achieve it.

Your goals need to be *measurable* so that you can track whether they are being achieved or not. As stated before, goals are broken down into tasks. This is the best way to determine if the goal is being achieved or if it is on its way to being achieved. It has the additional bonus of making it all look less daunting, as you can focus on reaching the smaller milestones in a shorter timeframe, confident that each achievement will culminate in the greater goal being accomplished. Using the healthier lifestyle example, if you want to lose 10 kilograms, and you have given yourself ten weeks to do it, then your smaller objectives might be to lose one kilogram per week. This will be the indisputable evidence that you are on your way to achieving your objective. You can refine your plan and your approach to achieving objectives if you find you are not on track, and you can approach them in a different manner. A colour chart indicating every task achieved within a goal is one way of tracking how well you are working towards your goals. For example, each gym session you complete successfully can be highlighted in green on an excel spreadsheet, which after a while will look like a sea of green. If you don't adhere to your program, for reasons within your control, highlight it in red. You will be able to track your progress over the weeks and months and will have the extra incentive of trying to keep a 'clean sheet' of green over the period of your program, knowing that every positive step brings you closer to your goals.

You have to be honest with yourself, and ask yourself if your goals are *attainable*. There's a difference between determining that something is not attainable because it's not possible given your current circumstances or that it's not attainable because you have told yourself that it's not simply because it's hard work. The former may change or improve over time, and you can set those goals for a later date when your situation improves. The latter simply requires you to ask yourself, 'How badly do I want it?' It's why you have to ask yourself, 'Why do I want to achieve it?' A military commander will conduct a feasibility study when setting military objectives. They will seek to determine if it is actually possible, with consideration to the time and resources currently available or required, as well as other constraints and limitations that they may or may not have control over.

Objectives may sometimes be altered to either overcome these constraints and limitations or to get around them. You need to do the same with your goals to ensure you don't set yourself up for failure by wasting time on something that really isn't achievable. Yet, you must also ensure that you don't miss out on an opportunity just because you didn't examine the ways in which you could have achieved them with a bit of effort. Look at the time you have available, which you have already determined in your own battle rhythm, and identify what resources or support you need to achieve your goals. Can you get them if you don't already have them? Is achieving these goals within your own capabilities?

The motto of the Special Air Service Regiment is: 'Who Dares Wins'. When you weigh up all the considerations, sometimes you have to be prepared to back yourself and think positively! If you critically, but optimistically, look at what you want to do and use this planning methodology, it will all appear less daunting, and you will start to believe more in your own ability to achieve things. We will look at the concepts of discipline and motivation in the second part of the book.

Goals must be *relevant*, meaning that you need to ask why you are undertaking the efforts to achieve this goal. Is it in line with your wider plans? Will it help you achieve the things you want to do? If not, you'll simply waste your time. Members of any Western military will do virtually anything asked of them. They have volunteered to join the military and have demonstrated a willingness to put their lives on the line for their country. The one question they will ask, however, is *why*. They don't want to believe they could be maimed or killed in a folly. Some of the greatest failures of the First World War were due to the ill-conceived and poorly executed operations that had absolutely no strategic objective. These were hard learnt lessons that came at the cost of hundreds of thousands of soldiers' lives. The modern military conducts carefully considered assessments as to how the military objectives are relevant to the wider plan, to ensure that valuable time and resources are not wasted unnecessarily. When planning your own goals, and the smaller tasks within them, you must take these considerations onboard, as wasted time is lost time. If something isn't necessary to the achievement of your wider plan, don't do it.

Your goals must be *time-bound*. Having an open-ended goal which you keep delaying is a goal that will never be achieved. Perhaps the greatest aspect military planners have to consider when conducting planning is time. There never seems to be enough of it. As stated in the introduction, time is a commodity, and a rare one at that. It will influence things more than anything, because time itself is inflexible (Wednesday comes after Tuesday whether we want it to or not), and deadlines often need to be met to ensure subsequent tasks can begin. 'Operation Jaywick' was a Second World War Australian commando raid into Singapore Harbour with the objective of sinking Japanese shipping. The special operations soldiers and sailors had to be released from their mothership, the *Krait*, at a designated time to allow them to paddle their canoes over long distances, to reach the harbour under the cover of darkness, so that they could place their mines on enemy shipping undetected. They then had to return to the *Krait* by a set time to allow all the raiders to make their escape back to Australia before the Japanese had figured out what had occurred once the mines had wreaked their havoc on their ships. The planners had to consider these time considerations when conducting their planning. As stated before, time is a commodity, and the time available to complete goals will have limitations. As such, you must set realistic time limits on your own goals, inclusive of any external influencing factors and based on your own capabilities. It is obvious to state that your goals could be achieved quicker if it was the only thing you were doing, but because most of us have many things we do in our lives, we have to take into consideration all these factors when planning. A timeframe will give you a date to work towards, but it must be realistic.

You need to make an honest assessment of how long you anticipate each goal will take, given all the other things you are currently doing or planning to do. This assessment doesn't have to be down to the minute, but should be enough to give you an understanding of how much time, based on your battle rhythm, you can allocate to each goal and, therefore, when you can expect to be able to complete it. It is possible to both overestimate and underestimate, so careful consideration should be given when deciding. You can determine this by looking at how long it has taken you to do similar things in the past. In the navy, tasks are called evolutions, and when briefing sailors to undertake them, the senior sailor or officer-in-charge will say (as an example), 'This evolution is expected to take thirty minutes'. This timeframe is based on having done that evolution many, many times before. Alternatively, you can break the goals down and determine how long each task within it is likely to take, then work out the cumulative total in time with these tasks. You should have time expectations on even doing the small things, such as cleaning your house. Failure to set these expectations will see your battle rhythm fall behind very quickly.

Finally, your goals must be *adaptable,* because your personal life situation might change rapidly and you need to be able to continue working towards your goals despite the changing situation. The principle of adaptability should apply to all the other principles too. For example, a goal to get fitter and eat healthier is harder to maintain over the Christmas period, when people are likely to spend more time eating and drinking and less time working out in the gym. It's likely that there will be social functions and family events that you need or want to attend, and this will inevitability take up time that would usually be devoted to being in the gym. You don't need to change your goals, but simply adapt them to the changing situation. Do this by ensuring that your goals aren't so restrictive or inflexible that a few changes in your life results in a complete inability to keep perusing them. Plan your goals with consideration that circumstances may change during the time you allocate to achieve them. During the 1982 Falklands War, the British plan to retake the islands involved having a large merchant vessel transfer to shore a large number of helicopters once the initial landings at San Carlos in the west of the island had been completed. This was to allow the Royal Marines and soldiers of the Parachute Regiment to be flown overland with the objective to capture the island capital of Port Stanley in the east. However, the ship carrying the helicopters, the merchant vessel *Atlantic Conveyor,* was sunk by two Argentine missiles, resulting in all but one of the helicopters being lost. As a result, the marines and paratroopers instead had to walk overland, a distance of nearly 100 kilometres, extending the timeframe originally allocated to meet the objective. The military objective was still met, because the British commanders adapted to the changing circumstances, and altered their plan accordingly. You must do the same. If your circumstances change, adjust your plan by again asking yourself the five basic questions and adhering to the principles previously established.

The military will always plan ahead, in some cases many, many years ahead (especially when looking at equipment procurement or recruitment targets) so that it can start to identify and undertake all the tasks it needs to do to achieve its objectives. This is called the 'strategic plan'. It is a broad outline of the things you need to do and want to do. It is broken down into specific parts, or as described before, tasks, with broad timeframes allocated to indicate when they will be undertaken. The biggest failure of the counter-insurgency effort in Iraq after 2003 was the initial failure to set specific, attainable and relevant military objectives, to set timeframes to them, and to adapt to the changing battlespace. Only after these principles were adhered to did the coalition start achieving success. Similarly, if you don't adhere to these principles, you will simply have a list of things you want to achieve that has no direction, just like a warship without a rudder.

From an individual planning point of view, you need to establish your own strategic plan, which becomes your personal plan. I recommend preparing a plan that looks out to four years from now. Many large corporate entities write their strategic plans at four-year increments. This timeframe is also the same time it takes to undertake a typical university degree. All commencing undergraduate students are told to make a plan that will allow them to achieve the type of degree they want to be awarded. They then look at the subjects they need to be studying in order to achieve that degree, and they prioritise the various courses and paths of study they need to undertake in order to achieve the degree.

Write your goals down and determine what you need to do to achieve them, then determine how long it is likely to take to achieve them (remember to adhere to the principles). Do this for the goals within each of your own focuses. This will take a little time, and you may end up moving some of the goals and tasks within them around depending on time available or personal preferences. Once done, you can start to look into planning when to start working on them. Divide it into a year-by-year basis for each focus, so that you have a list of all the goals you want to achieve, by focus, each year. When you first set up your own C2 Framework, you will have to decide which goals you want to achieve first. Some will be by choice, others because they are the prerequisite for moving onto the next goal. For example, your goal might be to buy a new car and start working as an uber driver. Your first goal might need to be to get your driver's licence, and then to save up the money to allow you to buy the new car. However, some things might only take a month or two, or can be done within a year, so you can do them in a shorter timeframe. Some focuses may not have any new goals in the later parts of the plan, and the goal here might simply be to keep doing the things you are doing within the focus (i.e. to keep a healthy lifestyle). If you wish to set a smaller timeframe for your strategic plan, then that is your decision. Once you have written it down, you need to sit back and look at it closely to ensure the goals are adhering to the principles.

Sometimes your goals may simply be enduring, and don't necessarily have an endstate (by where it's finalised, i.e. the completion of a degree). For example, one goal may be to maintain a good fitness base, by doing the same routine that you have been doing for years. You still need to allocate the time to undertake your physical training and to have a way to measure its effectiveness. The planning specifics will be outlined in the next part of the book: 'Current Focus—Planning it'. This is where we look at allocating time to actually undertaking the tasks needed to achieve your goals.

SUMMARY POINTS

- **Goals are achieved through proper planning and time management.**
- **Proper planning will allow you to allocate time to set out and achieve your goals.**
- **Ask yourself:**
  - What do I want to achieve?
  - Why do I want to achieve it?
  - What do I need to do to achieve it?
  - When am I going to achieve it?
  - What will it look like at the end? (How do you measure success? When is the goal achieved?)
- **Stick to the principles (SMARTA):**
  - Specific
  - Measurable
  - Attainable
  - Relevant
  - Time-bound
  - Adaptable
- **Create your own strategic plan, and consider what you will do and when (realistic time expectation).**

WRITE DOWN YOUR GOALS
(DO IT BY FOCUS and BY YEAR)

Current Focus—Planning it

"If you fail to plan, you are planning to fail"

BENJAMIN FRANKLIN

Everyone has their own future plans and intentions, some are just more considered and detailed than others. People say to themselves, 'In three years, I'm going to take a holiday' or, 'I'll finish my master's in the next two years'. Some plans are large and ambitious, while some are very basic and easily achieved. Your plans should be tailored to your own desires and needs. As stated in the introduction, the key reason that plans fail is because people don't properly determine how they are actually going to achieve their goals. Failure occurs when you don't set milestones or timeframes in which to achieve your goals.

I have explained focuses previously so we have an idea of what things we devote time to in our lives. We have looked at goals to articulate what it is we want to achieve. We have then divided our goals by each of the focuses. Additionally, we have looked at the battle rhythm to determine how we spend our time on a daily and weekly basis so that we can determine how much time we have available to undertake the tasks required to achieve our goals.

'**Current Focus**' prioritises what (and when) you are going to work on in the short term, so you can plan your days and weeks in order to achieve the goals. As mentioned in the previous chapter, I recommended having a strategic plan over a four-year period. Current focus will break that down into months within a single year, so you can articulate (and, therefore, know) more succinctly what it is you need to do and when. Current focus is the pivotal part of the C2 Framework as it's where you get to 'pull the trigger'. This means that up until now, everything has been put down on paper as an expression of intent. Now is when you need to determine when to actually do it. Figure 5 outlines 'Current Focus' within the C2 Framework.

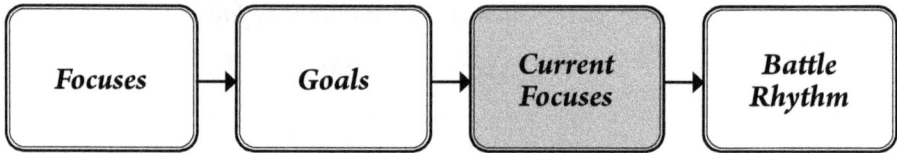

```
┌──────────┐   ┌──────────┐   ┌──────────┐   ┌──────────┐
│          │   │          │   │ Current  │   │ Battle   │
│ Focuses  │──▶│  Goals   │──▶│ Focuses  │──▶│ Rhythm   │
│          │   │          │   │          │   │          │
└──────────┘   └──────────┘   └──────────┘   └──────────┘
```

Figure 5—Current Focus

Current focus is where you look at your strategic plan and start planning to insert your goals, and the tasks within them, next to the actual months and days that you will do the things required to achieve them. This is known as operational planning. When conducting operational planning, a military commander must decide when, where and how they will actually devote the available time and resources to achieve their objectives. This is the same process used when looking to prioritise your goals over the duration of your strategic plan. As explained when discussing goal setting, you need to assess how long it will likely take you to achieve each goal (or task), based on the length of the task and the time available as determined by your battle rhythm.

From here, you can insert each of the tasks within your goals into your battle rhythm, having also considered two additional things, 'prioritising your decisive events' and your 'main effort'. Let's look at these two concepts.

You must remember to think about what tasks need to be undertaken, and in what order, to achieve your goals most efficiently. This is called 'prioritising your decisive events'. Think back to the example of the Normandy landings discussed in an earlier chapter. There were numerous decisive events that had to be achieved before the overall objective could be reached and the operation considered a success. In your own case, it may be as simple as needing to join the gym before you can start training and achieve your goal of losing 10 kilograms. Becoming a member of the gym is a prerequisite to gaining access to the treadmills and commencing your gym program. A more complicated example may be that you need to complete your tertiary degree before you can attempt to get a certain job, as employment at that organisation might be dependent on having that qualification. These decisive events will be pretty simple to determine for each goal within each focus. You then need to determine which order to place them and how long they will take to achieve. Do this by considering when they need to be achieved by, then work chronologically backwards so you know when you need to start them. They must be outlined accurately so you don't waste time doing things in the wrong order. Some goals may only take a few months to achieve, some may take years.

Even with a large army, the largest in Europe at the time, Napoleon had to consider where best to devote his soldiers and equipment when planning and conducting numerous battles. This was to ensure he didn't spread his forces too thinly on the ground, and thus waste valuable time and resources. He had to decide where to put his 'main effort'. Remember, time is a commodity, you only have so much of it, so you have to decide where to put your main effort when conducting your planning and allocating your time. Current focus means just that, what you are focusing on (working on) currently at a given point or period of time. This will be influenced by personal preferences or by the inherent requirements of your strategic plan. When considering factors, such as time available, you must determine what tasks and goals you *want* to achieve first and what goals you *need* to achieve first. For example, if you want to lose that 10 kilograms by Christmas, and it's only two months away, then you would have to start working on that goal straight away, while starting an online study course may have to wait until a later time because your schedule is already full.

You will have goals within each of your focuses, so you need to plan your time with consideration of needing to work towards multiple goals concurrently. However, sometimes you may find yourself fully devoted to only one focus (and the goals within them). It's not uncommon for students at exam time to ease up on other activities in their lives to devote more time to studying for the final assessment. This is the 'main effort'. Sometimes the main effort is spread across several goals, and you can find yourself working on multiple tasks because none of them are urgent or separately need to have more time devoted to them. This is possible, especially if you have set your battle rhythm so that you spend some time each day on each of your focuses. By having compiled a strategic plan, you will be able to work out when you need to focus on certain tasks more than others in order to achieve multiple accomplishments.

The military uses some fairly detailed techniques to conduct its operational planning. The larger the activity, obviously the more detailed the planning is as there are more aspects within the plan that need to be considered (known as the 'moving parts'). One of the most common approaches is the 'military appreciation process' (MAP), used predominantly by the Australian Army. The US military uses a similar methodology, called the 'military decision-making process' (MDMP). These are both stringent and well-outlined procedures where an individual or team of planners will examine closely what their mission and objective (goal) is. They then determine all the things that they need to do to achieve it, consider the amount of time it is expected to take and then look at how it can be achieved. Typically, several options to undertake and achieve the task are considered, and then the best option is chosen by the commander.

During this process, the planners must take into consideration all the moving parts that come together, as there might be multiple objectives to be achieved. For example, what will the naval forces be doing while the land forces are moving around? How does that interact with the air elements? How will they all be integrated? This is called 'de-confliction'. It ensures that all the resources and time have been allocated properly, and that they aren't trying to impossibly achieve two things simultaneously. Additionally, a process called 'wargaming' will occur. During this process, intelligence professionals will look to find faults in the plan and anticipate unforeseen events, which allows the planners to improve it accordingly. A commander gets the smartest people in the unit, who weren't involved in the original planning, to deliberately find holes in the plan and to identify what the planners might have missed or not considered that could make the plan fail (which can happen either due to bias—known as groupthink—or from the blinkered effect that can occur from spending lots of times in maps and paperwork). This process is known in US circles as *Red Teaming*.

As an individual, you can only do one thing at any given moment in time. You want to ensure you haven't planned to do your gym session at the same time that you are supposed to be giving that brief to your boss. You can't be off on a holiday when that essential week-long work seminar is happening. You need to do your own de-confliction and wargaming, to ensure you don't waste your time pursuing too many things (although keep in mind that on any given day or week you might have multiple tasks over multiple focuses). You must consider where your plan could go wrong and take steps to prepare for those potential eventualities. This is why you have focuses, so when putting the plan together, you can look at all the goals within them in an organised manner, and consider how they can all be integrated into one single plan.

To conduct de-confliction and wargaming, you will need to sit back and think about where you will be in a week, a month or even a year, and consider what things might possibly come up in the future that will impact on your plan. This is called forecasting. What will you do if you don't get that promotion you are working towards? What happens if you get injured during the ten-week training program? What happens if you get sick and are in bed for a week? Run it by family, friends or colleagues if needed, as a fresh pair of eyes may see something you haven't. Planning is supposed to be simple, but as the saying goes, 'it's possible to miss the forest for all the trees'! Plan for these risks by ensuring that your goals have adhered to the SMARTA principles, so that if changes need to be made, you won't have to completely write-off your goals, but at worst delay them and rearrange when you will do them.

This methodology, with a few variations, is used to plan military operations ranging from an entire combat campaign involving hundreds of thousands of troops, right down to how a thirty-person strong combat support platoon will tactically transport aviation fuel to a remote airfield. Military plans, when written out, can be pages and pages of words, charts, maps and other useful information. You don't have to have your plan that complicated or lengthy though.

PUTTING YOUR PLAN INTO THE BATTLE RHYTHM

Now is time to put all these tasks and goals, fully prioritised, de-conflicted and wargamed, into your battle rhythm, so you can start achieving your goals. The current focus box (or list) will detail all the things you want to achieve in the current year (remember to adhere to the principles of goal setting as listed in the earlier part of the book when determining where to put them). However, they should still be divided by focuses to allow you to determine what is coming up in your schedule as part of your overall plan. The month you are currently in is where you focus specifically on the tasks at hand, confident that every small achievement is one step closer to your goals, and that you are using your time efficiently.

This is the epitome of current focus, where you assign your tasks into your battle rhythm, having been prioritised by your decisive events and main effort. It will determine what you do each day, week or month. Over different days, weeks and months, you may find yourself doing more work towards some focuses compared to others. This is simply a reflection of what your priorities are. Naturally, there will be the tasks which are a consistent part of your battle rhythm, such as paying your bills or preparing your meals for the week. It is what you are giving your attention and time to at that particular moment (hence the term, current focus). Because your planning will have been well thought out having used this methodology, you can completely focus on the task at hand, assured that you are not forgetting anything.

If your focuses are small in number, and your goals have similar timeframes (or are enduring), then you might choose to have spaces in your battle rhythm evenly devoted to your focuses. For example, Tuesday evening might be devoted to hobbies, Wednesday evening might be devoted to the gym. This is a good way to avoid getting bored from working on the same thing constantly, and can prevent you from getting fatigued and losing motivation. It also ensures that you are progressing your goals because you have devoted time to work on each of them.

Write it out and track it anyway you like. An excel spreadsheet can outline the focuses in the vertical columns, the years in the horizontal columns, with the goals placed in the corresponding cells. A list on an electronic word document might work for you, or even writing it out by hand on paper. It's entirely up to you. Remember, it's supposed to be easily managed and easily referenced whenever you look at it. It's your plan, do what works best for you. The visual C2 Framework that I have utilised throughout the book might be sufficient enough, as you can fill in the boxes. A flow chart along those lines using many colours to indicate each focus, goal and year might be easy to manage (an example is in Figure 6).

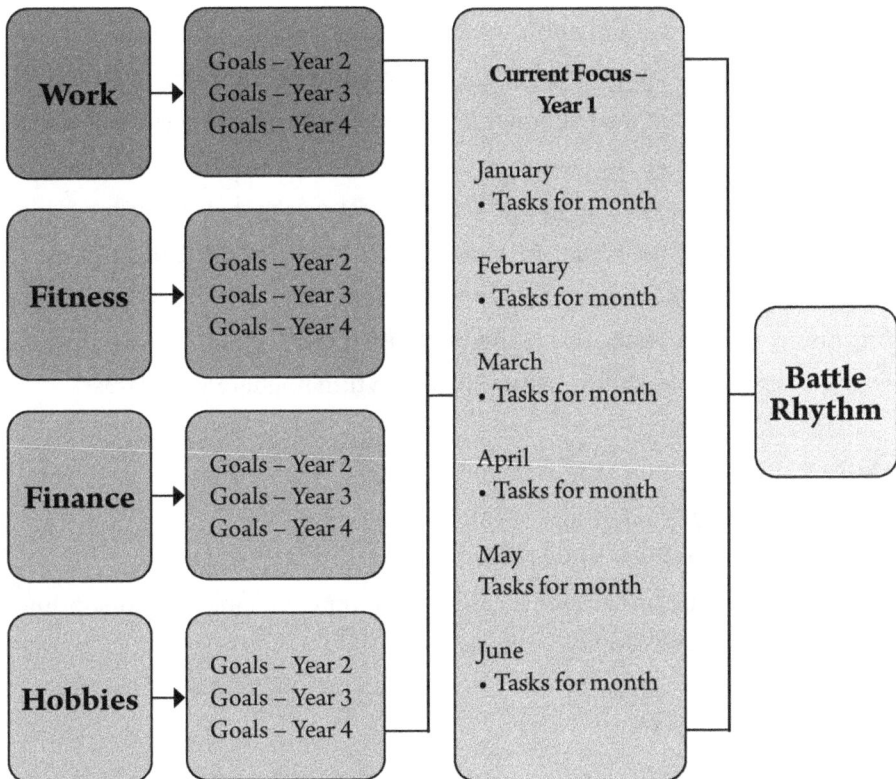

Figure 6—Example of one possible planning and tracking tool

You would then enter this into your diary, your calendar or even onto your bathroom mirror (there is more than one way to do it), in whatever way that best suits your own style, so that you now know what you want to be doing each day over the next month. As stated before, you don't need to run a battle rhythm that is planned down to the minute. Having a battle rhythm linked to a plan, even a basic plan, ensures you are making time to do the things you need to and want to in life. It might be as simple as allocating a day to paint the bathroom, and another to mow the lawn, knowing that you have remembered and made time to take the kids to sports and to visit the dentist.

Although I highly recommend having a daily battle rhythm, it doesn't make too much of a difference whether you plan a battle rhythm that works a day at a time, a week at a time, or a month at a time. You still have your current focuses written down for the whole year, so you can insert them into a battle rhythm that suits you. If your battle rhythm is set out on a weekly basis, it will still include items you only do monthly or annually (e.g. seeing the dentist). Now that we have looked at all of the C2 Framework, look again at the battle rhythm you started at the beginning of the book, and finalise one that will best suit you and your own requirements based on what your goals and focuses are. *It doesn't have to be rigid!* If you prefer to have a weekly battle rhythm with a list of all the things you want to achieve, rather than breaking it down into a daily battle rhythm, then do that. The C2 Framework is designed to help you organise your time better and help give you direction towards achieving your goals. An example of a fully inclusive battle rhythm might look like this:

Daily Routine

0600:   Wake-up, stretching session
0700:   Breakfast, leave to work
0815:   Work (one hour lunch at 1200—do some personal administration)
1700:   Finish work, go to gymnasium
1800:   Dinner
1900:   'Evening activities'
2100:   Review day and plan for tomorrow
2115:   Read book (thirty minutes), then bed

Evening Activities

Monday: Basketball match (no gym today, game starts at 1830, dinner after)

Tuesday: Study for TAFE course

Wednesday: Free time

Thursday: Spare evening to do what I like

Friday: Drinks with friends

Monthly

- Review achievements for month and plan for next one (at end of each month)
- Review personal cash expenditure and determine if savings plan is on track
- Time self in 10-kilometre run

Annually

- See dentist (February)
- Talk to bank about home loan options (July)
- Allocate goals for next year (November)

Not everyone needs to run a battle rhythm like Elon Musk or the US president. It will really depend on how busy or quiet your life is. When you are first using the C2 Framework, you will ultimately end up tinkering with your battle rhythm to find what works for you. The military will always plan. But often the best plans are improvements on previous plans and are the result of trial and error. So be prepared to learn from your mistakes. The success of the Normandy landings was because of the lessons learnt, very difficultly, from a failed attempt to capture a French port at Dieppe in 1942. This failed attempt resulted in many Allied deaths, with many more captured by German forces and placed in prisoner-of-war camps. The mistakes were examined in critical depth, and an improved way to do the same thing was developed, which led to the successful D-Day landings two years later.

SUMMARY POINTS

- Break your goals down into monthly blocks, or in a way that best suits you.
- Prioritise your requirements to achieve each goal (decisive events).
- Determine your priority tasks and goals for each month (main effort).
- Conduct de-confliction and wargaming.
- Insert your tasks into your battle rhythm.
- Start being more efficient in your time management and personal planning!

C2 Review

"Retreat? Hell, we just got here"

CAPTAIN LLOYD WILLIAMS, USMC

The purpose of the C2 Framework is to allow for better personal planning and time management. It is designed so you can articulate what is important in your life, set goals associated with them and establish a routine which efficiently allows you to manage your time in order to achieve those goals. It utilises well-defined military planning skills to manage your own life in a more efficient way. The result should be a more focused approach to doing things you want to do and with more time available to actually do them.

The planning methodology can be used for anything you need to plan. For example, if you are planning a holiday, first determine what you want to do on the holiday, how much time you will have, how to best fit in all you want to do and what you will need to take with you on the holiday. You then set about planning and preparing to enjoy your holiday, because you have considered all the factors. This ensures you haven't left anything until the last minute, and you can enjoy your holiday without wondering if you left your favourite pair of boardshorts at home!

REVIEWS

When the military conducts planning and operations, it ensures that these plans are constantly reviewed so that it can adapt to any changing circumstances and to assess the overall effectiveness of the plan. This then allows the military to manage the next part of the plan when the time comes to implement it. This is called an 'after-action review'. Similarly, you must make a time within your battle rhythm to ensure you can conduct reviews on your own plans, to measure the progress of them. This allows you to make changes as needed, and it is the opportunity for you to critically assess the effectiveness of your plans as you undertake the tasks within them.

A review will look at all the goals being worked towards within each of your focuses. It may be reviewing your weight loss program to critically look at what you did well in the gym, and determining what you could have (should have, perhaps!?) done better. It might be to review your cash expenditure for the month if you are trying to save money. Even relationships get reviewed to ensure everyone is getting out of it what they want. A review allows you to work out what you did well, what you could have done better, and what you need to do to get the plan back on track if needed.

Don't avoid doing this!

Take the time to review your plan. Regardless of how hectic or quiet your life is, you should always take a moment to pause, take a deep breath, and examine how you are going with life. The sheer inertia of the modern world sometimes means we forget to do this, yet it's vital to take these moments out to reassess. It is possible to get so caught up in your plans that you forget why you set out to do them in the first place. A review lets you re-examine your motivations and instil confidence in yourself when you see that you are actually progressing your goals and are still on track to achieving them.

You should not spend an inordinate amount of time conducting a review. However, the use of a small amount of time will prove well spent as you will be able to assess if your plans are still on track, and this will prevent you from inadvertently wasting time by going off on tangents with your plans, which can happen over time. The reason naval ships at sea take bearings of their location at frequent intervals is because, without doing so, over time those small deviations off course could result in ending up completely in the wrong place.

The shorter the timeframe you are reviewing, the less time you will need to devote to it. At a minimum, you should conduct reviews in line with how you have established and run your battle rhythm. For example, if you run a daily battle rhythm, then at the completion of each day you should conduct a quick review of how well you adhered to it, what you did and didn't achieve, and update, adjust or confirm it for tomorrow. It should be a simple case of adding to your diary the tasks you will do the next day (if they're not already in there). The whole thing will probably only take a few minutes. Most days should begin the night before (when you go to bed). Make an outline of what tomorrow will involve so you know when you need to get out of bed and what tasks you'll be doing for the day.

If you also have a weekly battle rhythm, then at the completion of that week you should review the execution of the plan and update, adjust or confirm it as required for the following week. For example, most working weeks commence on Monday, so a quick review on Sunday evening will mean you are in a good position on Monday morning, already prepared to commence the week. It's a simple check to ensure that you are staying on course and should be part of your battle rhythm. Do the same for a battle rhythm that includes events at monthly and yearly intervals. I recommend making some notes weekly so that you can clearly define what things you did well in the previous week, and what it is you want to do better the next week, to help keep your plan on track. It might be a case of wanting to be better at meeting your deadlines at work, spending less time in the gym procrastinating or devoting more time to a personal hobby. You can then use these notes during the week to remind yourself of the areas you need to be better at.

A review of your current focus, if you are running it on a monthly basis, should occur just before the end of the month. This allows you to start the first day of the new month prepared to continue the existing tasks and start the new ones, fully confident that your plan has been adjusted and that it still meets your needs. A full review of your year should be done just before you commence the start of a new one (whether you measure it by financial or calendar year), so that you can start putting the upcoming year's goals into your current focus. Try not to leave it until midnight on 31 December to start thinking about it!

By inserting these reviews into your battle rhythm, they will automatically become part of your routine, just as much as your morning coffee and shower. This will become second nature, as will some other aspects of your battle rhythm because they are likely to become fairly repetitive. However, even if you have a very dynamic battle rhythm with many changing facets, the reviews will assist you to maintain a planning cycle conducive to your needs, without taking up significant amounts of your time. However, avoid falling into the trap of reviewing a plan but then doing nothing about it. You must act upon the changes needed. An army commander will review old plans when making new ones, as they are a valuable source of working out what did and didn't work. Without making these adaptations, you'll simply waste more time on repeating the same mistakes, so that is why you should take notes when doing reviews, particularly if things didn't work well within your plan.

A review will also give you the time to conduct initial or updated planning (i.e. 'plan to plan'). For example, if you want to change your approach to your fitness routine to work towards your larger fitness goal, then you might need to develop a new six-week program. Another example would be the start of a new semester at university. You may need to make some adjustments to your timings and routines so you can approach the new semester differently and that you are still allocating the correct amount of time to your studies. These small adjustments in your focuses, goals and battle rhythms need to be conducted as required, so always make the time to do it as needed. Don't forget that you should even plan to make time for the smaller things that are an infrequent part of your routine, such as buying Christmas presents for friends and family. Thinking ahead to upcoming events, and determining when to do the tasks required of them and how much time they need, will ensure you are not trying to find time at the last minute to complete them.

When using the C2 Framework for the first time, your first review is the establishment of your own framework. This is where you look at and set your own focuses, goals and battle rhythm. Maintain the things that you already do well, and look to improve or fix the areas that you are unhappy with by using the C2 Framework to give more organisation and structure to your life.

C2 FRAMEWORK—SUMMARY

The C2 Framework is designed to be an easily managed personal planning and time management tool. It is based off some very well-defined and proven military planning and time management methodologies. Anyone can utilise it—CEOs, shop assistants, students—to improve their personal planning and time management. It will, at the very least, allow you to take stock of how you use your time. From there, you can decide to either add some things to your schedule, or if some things are preventing you from achieving the results you really want to achieve, then you can reduce the time you devote to them or remove them from your schedule completely.

Take the opportunity to apply the C2 Framework to your workplace. Review the battle rhythm of your workday to see if your workplace is wasting time. Use the methodology for managing projects and tasks. The 'Focuses' will be what your workplace does, the 'Goals' will be the outputs, whether they are changing or constant. The 'Current Focuses' will be the management of these goals (essentially project management). Tweak it to suit the requirements of your workplace.

A common saying in any military is 'Keep It Simple, Stupid'. It is known as the *KISS* principle. The more complicated a plan, the less adaptable it is and the more likely it is going to fail. Do the same with your planning, always keep it simple. It's not a complicated methodology. Change it as you need. For example, your battle rhythm will likely change several times … allow for that. Once you have achieved a goal, you set a new one. The measure of success is if you are achieving your goals and using your time in a more efficient way. If you only use part of this planning methodology, then that is entirely up to you. Entire battle-groups, with hundreds of thousands of soldiers, are constructed to meet the needs of a specific mission. So, you need to establish a plan, and a way of using it, that suits your own needs. Be accountable when reviewing your plans to see if you are succeeding in working towards achieving them. If not, be prepared to make changes. Telling yourself that 'all is good' will only result in failure. Be prepared to have other people look at it if you are unsure whether your plan will work. They can help with your wargaming.

Do not over-plan! Even the best military planners can get bogged down in overly complicated plans. This is a result of not adhering to the *KISS* principle, and trying to account for every possible scenario, to the point where all the resources are unevenly distributed and no decisions are actually made. As General George S. Patton said, 'a good plan violently executed now is better than a perfect plan executed next week'. That's not an excuse to come up with a half-baked plan, but rather to show that once you have a plan created and thought through, you should start to undertake it, and you can adapt it and amend it as you go along.

Staying disciplined and motivated can be the hardest part of achieving your goals. As mentioned in the introduction, great plans are for nothing if you are not disciplined enough to do the things required of them. The next part of the book will look at how the military instils discipline and motivation into its people, and how this discipline and motivation has allowed them to achieve significant victories despite considerably adverse situations. From here, I will show you how to utilise some of these skills in your own life.

C2 FRAMEWORK CHECKLIST

- **Establish your 'Battle Rhythm' based on your own needs.**
- **Look at and determine your 'Focuses'.**
- **Determine your 'Goals' (what do you want to achieve) and ensure they adhere to the 'SMARTA' principles.**
- **Prioritise those goals and establish your strategic plan with a suitable timeframe.**
- **Break these down into 'Current Focuses' and ensure you:**
  - Prioritise your *decisive events*
  - Determine your *main effort*
  - Conduct *de-confliction* and *wargaming*
- **Re-examine your battle rhythm and ensure it is:**
  - Flexible
  - Practical
- **Make time to conduct reviews (these are critical and should not be avoided!)**
- **Don't over-plan!**
- **Start being more efficient in your personal planning and time management! GET OUT THERE AND DO IT!**

PART 2
DISCIPLINE AND MOTIVATION

Discipline

*"The truth of the matter is that you always know the right thing to do.
The hard part is doing it"*

GENERAL NORMAN SCHWARZKOPF

Discipline

verb

1. train (someone) to obey rules or a code of behaviour;
2. train oneself to do something in a controlled and habitual way.

We will look at both discipline and motivation in this part of the book. Motivation is a positive enabler, it is a great thing to have because it gives you the drive and inspiration to do and achieve something, perhaps well beyond what you thought was possible. However, motivation can be fleeting, especially if your goals take a long time to accomplish. Without motivation, you can still pursue your goals and work hard to complete the tasks that will allow you to reach them. How? You do this by having discipline. Without discipline, it is very hard to achieve your goals. You may obtain a certain measure of success. You may even make it most of the way towards those goals and get to a point where you convince yourself close enough is good enough, but it will never be considered 'mission success'. Any person with integrity will know that they haven't achieved the outcomes they initially set out to do. It's by having discipline that you can keep chasing your goals, in an efficient manner, even when motivation is low or completely gone.

The military is renowned for its discipline. There is even a common phrase, 'military-like discipline', which conjures images of an individual or an organisation conducting itself in a manner expected of well-trained soldiers. Discipline is the core basis of how a military operates. Why? Because the very nature of uniformed service requires individuals to follow orders and act as part of a team even when working with conflicting personalities, both of which require them to deport themselves in a manner that demands significant self-restraint. It is critical to be able to be successful during peace and war because, if a military body is not cohesive, it is unable to work as a team and will not be able to undertake the highly complex and dangerous tasks that characterise military service. As Lieutenant-General Mark Hertling (former Commander US forces—Europe) discussed with *The Washington Post* in 2011, any lack of discipline in a military is considered 'cancerous', as a lack of discipline among a few can quickly turn into a lack of discipline among many, and will severely undermine the effectiveness of a fighting force.

Discipline is instilled into soldiers the minute they sign on the dotted line to enlist. It is taught using a myriad of techniques. Each nation's armed forces have their own methods of teaching discipline (some much harsher than others), but there are many common traits. The use of parade drill (the marching up and down on a parade ground) teaches soldiers to follow instructions and learn a basic level of resilience in order to achieve a required outcome. When either standing 'to attention' or 'at ease', a soldier is required to resist the urge to scratch their nose or brush away the fly that is crawling into their ear. This foundational teaching allows soldiers to develop a high level of self-restraint, character and efficiency, despite the urge to make themselves more comfortable or take the easier option. This is further enhanced in subsequent training, and then on operations, where they may be required to walk silently through dense and inhospitable jungle, as opposed to walking in the open and on the developed road, because it provides the tactical advantage required of the mission.

The military instils discipline through repetition and punishment. It trains its members to do things repeatedly so that it becomes second nature, known as 'the norm'. In the days of Admiral Lord Nelson, breaches of discipline by naval personnel were punished with the cat o'nine tails (a multi-tailed whip used for flogging errant sailors). Discipline was necessarily tight, because the life of a sailor was hard, and a failure to maintain discipline during periods of long inactivity at sea could result in mutiny, by both officers and sailors (think of the mutiny of the HMS *Bounty*[1]). However, it had a more specific purpose. Discipline was instilled, maintained and enforced to ensure that when the ship was called upon to engage in combat, all the officers and sailors knew their specific role. Despite every individual urge there might have been to run away, the sailors overcame their natural fears and remained focused on their job to ensure success in action.

It was similar for soldiers of that same period, who were forced to maintain straight lines abreast when facing the enemy as they marched into the field of battle, despite volleys of gunfire being directed straight at them. The discipline required of these soldiers to carry out their duty was immense, but it was needed to ensure that the battles could be fought and won in the manner that they were during that period of history. Although these tactics were eventually proved inefficient and foolish during the slaughter that was the First World War, a change in military tactics over time has not negated the need for those who perform them to remain disciplined when engaging in combat.

All Australian soldiers enlisting into the army go through recruit school in regional Australia at a place known as Kapooka. Discipline is taught through parade drill, as well as through many hours spent in the

---

[1] On 28 April 1789 on the HMS Bounty, a group of Royal Navy sailors, led by Lieutenant Fletcher Christian, placed Captain William Bligh and his supporters in a small boat and set them adrift in the South Pacific. In an extraordinary feat of maritime navigation, Bligh managed to use his skills in astronomy to navigate the small boat 3500 nautical miles to safety.

bush conducting training activities, which, depending on the time of year, can be in the extreme cold or extreme heat. However, one of the most basic techniques the army uses to teach discipline is known as the morning routine. All recruits are woken at 0600 by 'Reveille', a musical piece played over loudspeakers across the base. From here, all recruits strip off their bed linen and stand in the doorway of their rooms (which stretch down the long hallway of their assigned dormitory—otherwise known as 'the lines') with these sheets over their shoulders. This is to prevent recruits, who might try to save a few minutes in the morning from having to make their beds, from sleeping on the floor at night. The recruits then have fifteen minutes to shave, make their beds and dress in the allocated uniform of the day, before appearing in three ranks (the army term for rows of people) outside their lines to be marched to breakfast. Failure to do so draws the wrath of the instructional staff who will punish those who do not meet the mandated requirements. The recruits will quickly develop self- and team-discipline, as everyone learns to work together to achieve the standards required of them. No one will want to let their teammates down and be the cause of the platoon doing fifty push-ups before breakfast!

The repetition of this routine over many weeks not only becomes the norm, it also becomes a mindset and a lifestyle. The adoption of this routine turns the former civilians into soldiers by demonstrating to them that they can do things under pressure and outside of their comfort zone. These lessons in being disciplined are lifelong lessons that remain with the soldiers throughout their entire military career. When the going gets tough, be it in or out of uniform, these soldiers will reflect back to those mornings in recruit school as the framework from which to draw upon when they are expected to demonstrate discipline in whatever they do. The training soldiers receive at recruit school teaches them that when things really become difficult, they have the internal fortitude to remain disciplined and achieve the outcomes they need, whether it be in a peacetime environment or during actual combat.

During the First Gulf War in 1991, the Royal Air Force (RAF) conducted numerous low-level attacks on Iraqi military installations. These attacks required them to approach their targets in a direct flight path in order to most effectively bomb them. These were the same tactics that had been utilised by RAF Bomber Command in the Second World War while attacking the German industrial heartland. Despite having the element of speed and surprise, the RAF pilots were required to travel through a considerable amount of anti-aircraft fire on the way into their targets, and even more on the way out once their presence had been noticed by Iraqi forces. Their tactics involved flying very close to the planes next to them in order to meet the objective of the attacks, which was the destruction of vital enemy communication systems and buildings without causing civilian casualties. They needed to show immense discipline (and some considerable courage) in order to conduct their mission.

A failure to demonstrate discipline in the military will result in the failure to achieve an objective, whether it be to complete training activities or conduct battles in extremely adversarial environments. Conversely, an extreme amount of discipline can see a military force overcome significant odds to achieve their objectives. For example, during the Battle of Long Tan on 18 August 1966, Delta Company (D Coy) of the Sixth Battalion, the Royal Australian Regiment (6 RAR), consisting of a little over 100 soldiers, held off a combined force of nearly 2,000 Viet Cong and North Vietnamese regular soldiers. For over three hours, during the afternoon in pouring rain, the soldiers of D Coy fought in a rubber plantation, close to their operating base in the Phuoc Tuy province, against a numerically superior opponent whose aim was to dislodge the Australian Task Force from their newly established operating base in South Vietnam.

Despite fighting a force far superior in numerical strength, the Australians, supported by New Zealand Army artillery, managed to defeat the attack and defend their base. The victory was achieved because the soldiers held their ground despite constant enemy fire, as their defensive positions were mutually supporting, meaning that each position had to ensure that the one to the left or the right of them was protected from any enemy attempts to outflank them. They also maintained their fire discipline, meaning that they fired their rifles only as required, to ensure they didn't waste ammunition. Additionally, they moved as needed to provide casualty management to their fellow soldiers, despite the artillery barrages landing very close to where they were fighting. The many days and weeks spent training, along with a significant amount of personal courage, ensured that this high level of discipline, essential to the success of any action, was maintained throughout the battle which saved many lives and ensured a victory was achieved.

THE DISCIPLINE TOOLBOX—FIVE WAYS TO BECOME MORE DISCIPLINED

We can draw upon these lessons so we can be disciplined in the way we conduct our own lives. As stated at the start of the book, great planning and time management means nothing if you don't have the discipline to actually do what is required to achieve your goals. There is one thing that all soldiers know—discipline is hard and there are no shortcuts. It requires patience, constant effort and stepping outside of your comfort zone. It is about doing what is right and what needs to be done when no one is looking, even though there might be an easier or more comfortable option available. There is one very important thing to keep in mind— *discipline is a choice* (and there is no way of getting around this). You have to force yourself to do the things you need to do. This may mean getting out of bed early if required, sacrificing the Sunday afternoon lounging around on the couch so you can prepare for your exam or going outside for your 10-kilometre run even in the dead of winter. Discipline is what makes a soldier stay awake while sitting in a water-filled gun pit (a foxhole) at 2 a.m., in freezing conditions, so that the rest of the platoon can sleep safely. It's what makes firefighters run up the stairs of a large, burning building to help get the inhabitants out. It's what makes a soldier jump out of a perfectly good plane using a parachute packed by a complete stranger. You can instil discipline in yourself, without the need to have a drill sergeant standing over you yelling instructions.

We used the C2 Framework to set up a routine that will enable us to achieve our goals through effective planning and time management. This in itself is the first step in developing self-discipline, as organised structure inherently supports discipline. However, it is only the execution of these plans that will see your goals achieved. There are various ways you can ensure you become, and remain, self-disciplined in order to achieve your goals and to improve your life generally. They will become your 'discipline toolbox'. These are the things you do to become and remain disciplined.

1. Take the Small Wins—Make Discipline a Habit!

To become disciplined, you need to develop enduring habits. In the C2 Framework, you break your goals down by focus so they appear less daunting. As you complete each smaller task, you not only track your progress, but you will feel a sense of achievement that each step is getting you closer to your goal. The shorter the duration of an activity, the easier it is to remain disciplined. The average human adult's attention span is twenty minutes (if Wikipedia is right!). Small wins in life, especially if achieved when discipline was required, can be inspirational enough to allow you to realise that being disciplined is something that you can consistently do. It allows you to remain disciplined when you start to fatigue, both physically and mentally, because the tasks you are undertaking are more difficult or take longer to achieve. You will know what a sense of achievement feels like and will, therefore, be less inclined to give up.

When soldiers are conducting forced pack marches over long distances, often many tens of kilometres, usually carrying packs weighing up to 60 kilograms, it is very tempting to want to slow down, drop the head and try to become more comfortable during the gruelling activity. Soldiers can't afford to do this. They need to keep their heads up, observe the environment around them and ensure tactical procedures are being followed. They do it to protect the soldier next to them and expect that that same soldier is doing the same for them, as an enemy could be spotted at any moment. They remain disciplined by breaking down the march into achievable milestones. One method taught to Australian infantrymen is to focus on the march 500 metres at a time, keeping their head up, being tactically safe, and keeping focused on getting to the next milestone (be it a tree, a rock or the next water stop). The same applies to other strenuous tasks. They break time down into blocks, and focus on getting through one hour, then the next, then the next, until the task is completed. They focus on the small wins that, when combined, allow them to remain disciplined over longer periods and to achieve the required outcomes. After doing this numerous times, these small efforts in being disciplined become the norm; they become second nature and much easier. When the time comes to show greater levels of discipline, especially for harder or longer tasks, the soldiers know that they have the capability to do it, and it becomes part of how they approach everything they do.

Making your bed first thing in the morning is a small, basic task, but it is one that can set you up for the rest of the day. As the soldiers at Kapooka quickly learnt, it's the first thing they do each day. They have now demonstrated to themselves that, regardless of the fact 'Reveille' was playing and instructional staff were likely yelling down the hallway, they can push through the difficulty of getting up when tired and do something that will allow them to start their day on a positive note. From here, they then can get into the routine of their training day. These habits stay with them forever, if they choose to maintain them.

You can also do the same! Break down your own tasks into smaller and more achievable milestones where possible, so that you can focus on and do them one by one. Parents teach this to their kids when they're procrastinating with cleaning their bedroom. The kids are told to tidy their bedroom a portion at a time, so that task seems less arduous. This technique can be applied to anything. When doing a long-term gym program, you can approach it a week at a time. When studying, you can approach an essay a page or chapter at a time. At the end of each part, take a moment to review how you went, then focus on and apply yourself to the next part of the task. Breaking tasks down and focusing on them a portion at a time will give you a small sense of achievement many times over, which can inspire you to seek out the next challenge and continue to work towards your goals. It will allow you to continually practice being disciplined because, unlike the soldiers at Kapooka, there is no one around to make you do it, so you need to rely on pushing yourself.

You can then draw on this practice when greater discipline is required to pursue the things you want and need to do. If you really want to challenge your self-discipline, take cold showers each morning. The Royal Australian Naval College (RANC) used to make the Midshipmen take cold showers when training at the college (not anymore though, which I was most grateful for when going through myself). In his memoir, *Breaking Ranks*, college graduate Peter Cabban discusses that even after graduating from RANC, he continued having cold showers for many years, even after he left the navy. It had become a habit, and a simple act of discipline which set him up for a productive day. For Peter, after forcing himself into a cold shower, everything else seemed far less difficult. By incorporating similar basic habits into your battle rhythm, you can give yourself some small wins at the start of, and throughout, your day, which will instil enduring discipline in your own routines.

2. Use What is Available to You—Think Smart!

Use anything at your disposal to assist you. If you want to get up early, set an alarm. Modern smartphones have multiple alarms, so you can even set several. That way, if you miss the first one, or hit the snooze straight away out of long-term habit, the subsequent ones will force you to stay awake. An additional option is to place the alarm away from your bed, so that you are forced to actually get up to turn it off. At this point you are up anyway, so you use this as an opportunity to begin your day. As you develop the discipline to get out of bed when the first alarm goes off, you won't need the extra alarms, as you will naturally just get up knowing that you have done the same thing many times before, and knowing that once you are up you can get a head start in beginning your day. Use those alarms to tell you when to start your next task—it can be like your own personal assistant. Similarly, if you need to get up early, and therefore should go to bed early, don't stay up late watching that movie. Record it or stream it later off of the internet.

Look at the resources for all your focuses, and what you could use to help build your discipline. Soldiers have a motto: 'Train hard, fight easy'. The way they do things is smart, not stupid. Don't try to do everything at once. The inherent nature of military service often requires one to carry out their duties despite serious discomforts. It still requires discipline to sit in a sniper overwatch position on the side of a snow-covered mountain when it's freezing cold. However, the sniper team won't do it while just wearing a flimsy cotton shirt and pants. They will still give themselves the best chance to be able to do the job, so they will ensure they have the correct cold weather clothing. If your fitness goal, such as training for a marathon, requires you to get up at 5 a.m. to go for a 10-kilometre run, then purchase a long sleeve running shirt to run in instead of a singlet. You still have to go out into the cold and you still have to go for the run.

However, the psychological advantage of having that extra layer over your arms cannot be underestimated. The actual act of going for a run will eventually become habitual, and you will likely still do it even when the dog decides to eat that long sleeve shirt one morning and you have to run in the singlet anyway. Choosing to run in the singlet will certainly build character, and many military training institutions choose to deliberately deprive students of even the smallest of luxuries in order to build resilience. US Navy SEALs undertake their basic training in the cold waters off of southern California, and often they will only wear issued green fatigues to work in, despite the abundance of wetsuits available. The Australian Army infantry school at Singleton is one of the coldest places in the country, and during the final field phase, infantry trainees are prohibited from wearing their cold weather gear when digging their gun pits in pouring rain after two consecutive days of no sleep. The soldiers learn to reach deep into their character in order to learn and maintain the discipline to keep going. Knowing they have a spare pair of dry socks in their packs waiting for them once the pits are dug is enough to maintain that discipline.

3. Be Brilliant at the Basics—Take the stairs!

Taking pride in everything you do means that you approach all tasks, regardless of how big or small they are, with the intention of doing them as well as you can. Military personnel wear a uniform not only as a form of identification, but as a daily means of maintaining discipline. Uniforms are regulated, often right down to how the boots are to be laced, how it should be ironed, and how many centimetres above, on or below the breast pocket the name tag needs to sit. This forces military members to become meticulous and well-practiced in all things they do, leading them to approach all endeavours in a disciplined and methodical manner. Soldiers take pride in wearing their uniforms, as it makes them feel that they are part of an organisation that is respected and worthy. They will habitually undertake the smallest tasks to a high degree of proficiency. They become 'brilliant at the basics'.

Taking pride in yourself and your appearance is uplifting. It can boost your confidence in approaching each day or activity, which will then help you to remain disciplined because you will feel mentally, and perhaps even physically, strong. Everyone knows, even if science hasn't explained yet, that having a nice haircut puts a little spring into your step. The same can be said about putting on make-up or having a fresh shave. Although you may not be part of a uniformed organisation, you could still wear a uniform of sorts to help you take pride in yourself and your appearance. When you put it on, your mind will automatically start thinking about getting into the activity it is associated with.

For example, if you work in a business environment, you may have a particular style that you wear consistently (e.g. dark coloured suits or a white shirt when giving an important brief). It doesn't have to be exactly the same every day, but it's a particular type of dress-style that you have made the effort to prepare and maintain, and it is your way of associating what you want to do when wearing it. When you put it on each day, you know it's 'game on'. The same goes for gym clothes where, like in the business environment, you could have a constant theme, such as knee-length shorts and a certain coloured top. Taking pride in these 'uniforms' gives a mental advantage that will help maintain discipline when undertaking the things you want to or need to do. It will give you confidence, as it is a simple act to put on your clothes, which will psychologically tell your mind that it's time to get into the gym, to get into your studies or to approach the boss and ask for a pay rise! Many ex-special forces and intelligence operators go into private business, with most choosing to wear what is known as 'the operator uniform'. This is typically a polo shirt, a rugged pair of cargo pants and durable sneakers (usually made by the footwear company Merrell). This is not only because it is often a requirement of the jobs they are doing, but because it is close in nature to the type of clothes they used to wear in the military. Wearing this 'uniform' gives them back that sense of purpose—an inherent requirement of discipline.

Being fit and active has benefits of being both mentally and physically stimulating. Soldiers undertake rigorous physical training as part of their job, to the point that it becomes a lifestyle. Many civilians do it too, and in recent years, there has been a flourish of social media sites devoted to physical fitness and wellbeing. It is far easier to learn and maintain discipline if you lead a healthy lifestyle. This doesn't mean you need to become a fitness junkie, spending two hours a day six days a week in the gym. It simply means looking after yourself so that you have the energy to do what you need to do.

We all get tired, and this is when we need to call on our discipline the most in order to finish that study paper or push through that intense gym session. However, this is far easier if you are not in a constant state of lethargy due to preventable poor health. Even if you lead a very hectic lifestyle, you can still find time to do some physical activity, which will give you that little bit of a post-workout high. If you live or work in a building with multiple levels, take the stairs as opposed to the lift. If you work on a very high floor, walk half way up, then reward yourself by hitching a ride in the elevator the rest of the way. Doing this several times a day will not only help instil discipline in you, but will improve your all-round physical fitness. Another method is by doing the 'push-up club'. This was something I often did with a friend—a very fit and intelligent officer whom I later served in a joint special operations unit with—when we were both in the navy. In push-up club, every hour on the hour during the working day we would stop what we were doing and drop to the ground and do a certain number of push-ups. This was not only great at instilling discipline, it was also a great way of improving physical strength. It might even inspire your family, friends and colleagues to do similar to help their own fitness goals. You will feel good about yourself and will be far less likely to take the lazy option when working towards your goals. It is practicing discipline on a daily basis, which helps it become second nature—the norm.

4. Stay Humble—Lose the Ego!

The most disciplined soldiers are almost always also the humblest ones. They are the ones without ego, who simply strive to get the job done, regardless of the difficulty surrounding the tasks they are undertaking. These are the soldiers who like to be referred to as 'the quiet professionals'. They have mastered the art of self-discipline, and they are often found in elite units such as special operations forces or airborne regiments. This is because these units have very difficult selection requirements, where candidates are pushed to their physical and mental limits in order to prove themselves capable of undertaking the training and subsequent job expectations required of serving in these environments. That said, being an exponent of self-discipline is down to the individual, and you can find very humble and capable people in all parts of the military.

Being humble means the only person you are trying to prove yourself to is you! It means knowing if you haven't done the things required by not being disciplined. Your harshest critic should be you, although you also need to be fair to yourself. For example, if you are at the gym and you haven't given your session 100%, you will know it. As discipline is about doing the right thing when no one is looking, you need to develop the habit of assessing your own efforts and pushing yourself when things get difficult or you get tired. By developing humility in all you do, you will never attempt to try to impress other people, but rather quietly achieve the things you want to achieve. The results will speak for themselves, and people will end up noticing anyway. Talking up a big game and bragging about your capabilities will result in losing the respect of those around you. It also means that you are likely the sort of person who cannot quietly go about their business and overcome difficulty when the going gets tough, because you have never had discipline instilled in you. Through learning and having it in your toolbox, you will be able to draw upon it when you need it most.

5. Remove Temptations—Jump on Green!

Although discipline is hard, it doesn't mean you have to maintain a gratuitous level of hardship. As humans, we are naturally designed to want to do things in an efficient way. Soldiers are not robots, they are still only human, that's what makes having discipline hard because it requires us to do things that are against our natural mental and physical tendencies. Soldiers deliberately choose to sleep in the rain even though they could have shelter. They choose to walk 80 kilometres even though they could drive in a vehicle. They do these things because the job of being a soldier requires it in order to gain the tactical advantage. It requires extensive discipline.

It's the choices we make in life that require discipline. As the saying goes, 'nothing worth doing is ever easy'. You could quite easily cruise through life doing the bare minimum, goodness knows there are many people that do.

However, most of us live in a world where hard work is rewarded, and we want to set goals and achieve things.

Give yourself some assistance by removing all the temptations and distractions that make it harder to remain disciplined. As stated previously, discipline is hard. There is no point making it harder for yourself unnecessarily. This holds especially true when you are first striving to instil a higher level of discipline in yourself. The first time you try to go on a healthy eating plan after a lifetime of eating poorly is a hard adjustment to make. Assist your own efforts by removing all the junk food from your fridge and cupboard. Just having that chocolate bar in the back of your fridge will be tempting as it will call out to you in the middle of the night when you start having hunger pangs. Turn off the television when studying so you don't get distracted. You will be surprised how much you will get done when only focused on one thing and not having the audio or visual distractions in the background. Nutrition is an integral part of any fitness plan, but it is usually the area where most people fail to adhere to their personal training programs. We've all been tempted to sneak to the shop for a greasy burger or to raid the shared fridge full of unhealthy temptations when living with people who aren't seeking to have the same diet you are. When tempted to deviate from your fitness plan, do something to distract yourself from that craving. Instead of taking a slice of the cake from last week's birthday party, quickly do a set of push-ups as a healthier alternative. It will distract you for long enough not to submit to the craving and will have the additional benefit of being a positive action working towards your fitness goals.

Conversely, occasionally reward yourself when you are consistent with your discipline. The Australian Army has a term called 'EKO'. It stands for 'early knock off', meaning a finish to scheduled activities earlier than the usual battle rhythm. This is typically given when soldiers have completed a strenuous working day or week or have completed a particularly difficult task when inside barracks. It has the psychological effect of giving soldiers something to work towards, knowing that their efforts will be recognised. Do the same for yourself. Perhaps head to a movie at the end of a hard study week, or enjoy a sleep-in on a Sunday after having made the effort to get up before sunrise every day during the week to get into the gym.

As you start to develop and improve your self-discipline, you will be better at overcoming these temptations. The good habits you have developed will mean that you won't be concerned about only setting one alarm to get up early, because you know you will get up when the first one goes off. However, you need to know your weaknesses, so you can approach your tasks with the appropriate levels of caution to ensure you remain disciplined. Some things are easier to remain disciplined with than others. You will need to determine in what areas that holds true for yourself so you can take measures to defeat those temptations that can lead to you becoming ill-disciplined. If you find it harder to get up early as the working week goes on, it may behove you to set those extra alarms on Friday, even if you didn't need them on Monday. Discipline is the ability to do difficult things despite having a desire, or reason, not to do it. Don't be perturbed if it often feels uncomfortable doing the things that require discipline. Discipline is a bit like courage, which Mark Twain defined as 'the overcoming of fear, not the absence of fear'. Discipline is overcoming the temptation to take the easy approach, not a lack of temptation to take it.

Paratroopers go through a series of preparations before they can jump out of the aircraft. Although they know when the drop-zone is approaching, they have no say in when they are to jump. They have to mentally and physically prepare themselves for the moment the doors open and the chalk (the group of paratroopers) stands up and prepares to jump. When the light goes green, they have to jump, whether they are ready or not. They have to be well-disciplined to not baulk at the door, even if they have mental reservations. Unfortunately, you often don't have a choice about when the things we want to do need to be done. You can't wait until you feel comfortable or when the situation is in your favour to undertake your tasks.

The War on Terror started unexpectedly when four hijacked airplanes wreaked havoc upon civilian and military targets in New York, Washington DC and Pennsylvania. The first responders on that day were not prepared to respond to such a massive number of casualties. Nonetheless, those brave first responders still ran up those stairs, and continued to run into the unfolding catastrophe to try to save lives even as buildings were collapsing around them. Similarly, the coalition formed in the aftermath of the attacks were, on the whole, not prepared to undertake unconventional counter-insurgency warfare due to years of having trained for large-scale conventional battles stemming from the Cold War. However, the attacks demanded an immediate response, and the militaries of the nations that responded to this act of terror had to adapt to the circumstances in front of them. As a result of strong discipline and resilience, they were able to adapt to a different type of warfare that they were required to undertake.

By developing the habits as listed within the toolbox, you will be able to deal with unexpected situations in a more disciplined manner and be better prepared to step outside your comfort zone to undertake the tasks that you need to achieve. You will inevitably do the things you need to do when they need doing. You will go for your run even when it's a colder morning than you expected. You will delve into that task at work even though the boss may have changed the deadline. You will always jump on green!

INSTILLING DISCIPLINE IN YOURSELF

Discipline is about being resilient and still working towards your goals even when doing so proves difficult. It also teaches patience, so that you don't make emotional or rash decisions without giving due consideration to something. It comes easier to some than others. You need to work out where your strengths and weaknesses lie, and always be prepared to learn, adapt and improve your own capabilities in regards to having discipline. Discipline applies to all your focuses, be it your time, your money, your workplace tasks or your fitness routines. Look at the areas where discipline comes easy, and where you really struggle to maintain it. Remember, there is nothing wrong with struggling to maintain discipline, it requires focus and a continual effort. It's by identifying the areas where you struggle that you can mitigate it by implementing the tools that assist you in maintaining discipline. Remove the temptations and distractions. It is about instilling the habits that make discipline a natural and automatic part of your approach to life. It takes discipline to constantly adhere to your battle rhythm; it takes discipline to pursue your goals, whether they be small or large; it takes discipline to be disciplined! It may not get easier. There will always be times where you really have to push yourself. In difficult times, when you really need to call upon your self-discipline, the best thing you can do is remind yourself why you are doing it. Always remember that there is an outcome you are working to achieve and that you *can* stay focused on working towards it.

Discipline can be practiced. Making your bed daily is just one way to instil it in yourself. Infantry soldiers condition their bodies for the rigours of arduous fieldwork by regularly going on the aforementioned pack marches, where they carry all their equipment in large backpacks as well as their weapons and other supporting equipment. A fully laden infantryman can carry in excess of 60 kilograms when they have all their necessary equipment. It's actually great exercise. However, even in the safe confines of their barracks, where the march routes are usually flat (or simply on the side of the sealed road and, therefore, easier to navigate), the soldiers will still treat these walks the same as if they were in an operating environment, holding their heads high, observing their surroundings, and not wearing headphones in their ears listening to music in order to kill the time. This is enforced discipline. Its effect is to ensure that when they have to do it for real, or when undertaking special operations forces selection which requires long individual pack marches sometimes in excess of 100 kilometres, they have the discipline within them to keep undertaking their tasks even when things start to become difficult.

By following these tips, you have started to ingrain the practices into your daily routine, such as making your bed daily or taking the stairs when available, which will allow you to face difficult tasks knowing that you have the ability to apply discipline when needed. This will also promote confidence within yourself regarding your own ability to achieve your goals, even though at times it is still painful or difficult. Self-discipline will become second nature—it becomes the norm, and you will have the ability to have discipline in all your endeavours.

SUMMARY POINTS

- **Military-inspired discipline can be used in your own life, and it enables you to achieve your goals.**
- **Get the small win, and make discipline a habit!**
- **Use what is available to you, and think smart!**
- **Be brilliant at the basics—take the stairs!**
- **Stay humble and lose your ego!**
- **Remove temptations—jump on green!**
- **Discipline teaches resilience and patience.**

WRITE DOWN WAYS YOU THINK YOU CAN BECOME DISCIPLINED

Motivation

"Because you can…"

ANY SOLDIER WHO HAS ENDURED WAR

Motivation

noun

1. A reason or reasons for acting or behaving in a particular way.

There are literally hundreds of books, articles, movies and quotes relating to motivation. A quick Google search will reveal the myriad of literature surrounding the topic. In a nutshell, motivation is what drives you to do something, it is the reason you do it. Motivation provides the willingness and inspiration to pursue your goals.

People elicit motivation from all sorts of sources. It might stem from love, from a desire for revenge, as a result of success or disappointment, from a specific event or from their own life circumstances and experiences. These are the things that drive and inspire people to do something that they may otherwise not have done. For example, watching a disabled athlete complete a marathon can motivate able-bodied people to get outside and start training, as seeing that person overcome a significant obstacle in order to complete a difficult task provides the impetus for them to strive to do something in their own life that they may not have otherwise undertaken.

We all have different things that motivate us. The very reason we choose our specific goals is the result of various motivations. A desire to hit the gym and lose 10 kilograms may be motivated by being unhappy with how you look in a swimsuit or from a desire to be able to play with your children in the park without losing breath. More seriously, seeing a loved one die from a potentially preventable illness, such as heart disease if resulting from years of unhealthy living, can be the motivation to start living a healthier life. Working hard at your job and seeking a promotion may be the result of the motivation to earn more money in order to save for a house deposit. Each of your focuses will have different motivations, or they might be linked. Most soldiers have 'work' and 'fitness' as a focus. They are inextricably linked because being a good soldier requires a high level of fitness. Their motivations are the same for both, to be the best soldier they can be, so they work hard on their soldiering skills as well as their fitness.

While discipline is the tool we use to continue pursuing our goals, motivation is what makes us want to set out to achieve those goals. There are two types of motivation—hard and soft.

Hard Motivations

Hard motivations are the things that *drive* you. These are the motivations that are more enduring and don't easily waiver despite the increasing difficulty of the activity you are undertaking. This is because these motivations come from a deeply held desire to achieve a goal, from a determination to overcome an obstacle or from the impact that a significant personal event had upon you. It is the very reason for why you are doing something. For example, after the September 11th terrorist attacks, thousands of young men and women from around the Western world sought to join the military in order to serve their countries. They were motivated to do something that was greater than themselves and their own personal ambitions, knowing full well the hardships that would likely lie ahead by participating in combat.

Another example is a struggling actor who is down to their last few dollars. They have experienced what it is like to have to spend each night on a different couch, having been rejected at yet another audition and spending long hours working menial jobs. Some choose to give up, while others, who often end up becoming the most successful actors in Hollywood, use these experiences of hardship as the driving force to take even the limited opportunities that are presented to them and to keep working hard to chase their dreams. They are motivated by the desire to become successful in their trade, to be able to say that they made it, despite everyone and everything suggesting that they are likely to fail.

Olympic athletes use the failure to win a race as the motivation to keep training for the next four years in order to be successful in the next Olympics. This is a strong and enduring motivation that makes them want to keep getting up before sunrise six days a week to train in the pool. This still requires discipline, but it is strongly assisted by a high level of hard motivation.

Soldiers join the army for various reasons. Usually a desire to serve their country, to get a pay cheque to support their family or as a means of escaping otherwise difficult life circumstances. These motivations help them endure the hardships of military service, and when trained in the art of discipline, combine to make them part of a professional force that can perform difficult tasks in incredibly hostile environments. The modern army, be it the United States or Australian, has achieved the successes they have in the past few decades due to the fact they are an all-volunteer force. Those that join have a desire to be there. Despite these motivations wavering at times, and almost all soldiers at some point will question their motivations for joining, there remains a certain level of hard motivation that assists them to continue performing their duties particularly during hardships. In the First World War, hundreds of thousands of young Australian men enlisted to fight in far off places as a means to serve 'King and Country'. When the realities of trench warfare soon exposed themselves on the Gallipoli Peninsula, and later on the Western Front, the original reasons for fighting were no longer relevant and their motivation changed rapidly. The soldiers began to fight to protect the man next to them so they could all get home safely. This has been a common theme through the history of warfare.

Special operations forces of various nations have very stringent selection requirements. Potential candidates are pushed to their mental and physical limitations for one reason only—to test their motivation to enter and serve in those units. The very nature of service in special operations forces requires highly motivated individuals, who can work in teams or autonomously, to achieve incredibly difficult objectives. These units only want people who can stay mission- and task-focused even when they are at their physical and mental limits. During the selection process, these conditions are simulated and replicated to induce these stressors, so as to ascertain whether candidates can operate within the special operations environment. The Australian 2nd Commando Regiment conducts an activity during their selection course called 'demarcation' where candidates are deprived of food and sleep, whilst undertaking tasks which encompass both rigorous physical activity and complex cognitive decision games, for several days. It forces the candidate soldiers to have to seriously ask themselves, 'How badly do I want it?' and 'Do I want to be a part of this organisation?' The US Navy SEALS undertake a similar well-known activity called 'Hell Week'. Their motivations (and discipline) have to be deeply manifested in order to keep undertaking the gruelling activities, as the motivation gained from watching a few war movies will not last long.

Examine and find your own hard motivations. They should be fairly obvious. It might be related to your health, your family, your financial situation or from an event that made you decide you wanted something better for yourself. You may not have hard motivations for everything you do. Your entire C2 Framework may be driven by only one hard motivation. That's fine. Know what it is and how it can shape your wider plans. It will assist you to keep pursuing your goals even after things start to get really difficult. When discipline starts to waiver, remind yourself of your hard motivation to remotivate yourself so you never give up on chasing your goals.

Soft Motivations

Soft motivations are the things that *inspire* you. They grab your attention and help boost your spirits. Soft motivations can also be the impetus for establishing new goals or resetting your focuses. They might turn into hard motivations later on. They provide the spark to want to do something, but will likely peter out and are far less enduring. This is because soft motivations are not the reason why you are doing something. They lack that personal affiliation that comes from hard motivations and are unlikely to drive you when the going really gets tough and you need to draw upon deeper, more entrenched motivations and discipline to pursue your goals.

Everyone is inspired by soft motivations. These motivations 'pump you up', giving you a mental or physical spark to want to undertake a task or activity with increased vigour. They might even encourage you to develop a new goal if you find it inspiring enough. Watching an inspiring movie, listening to an inspiring story or seeing an inspiring action, regardless of what it is about, can motivate you to work a little harder towards your goals, to step out of your comfort zone or simply do more with your life.

Sometimes, soft motivations will relate to what you are trying to achieve, for example, when working on your fitness or health goals. YouTube is full of fitness videos that are great to watch, and they usually make you strut towards the gym with a renewed enthusiasm for your next session. Watching montages from the *Rocky* series is always motivating. However, although inspiring to watch, unless your goal is to be superheavy-weight champion of the world, it is a soft motivation that will fade very quickly from memory when you are in the middle of a particularly hard cardio circuit, trying to remember why you are in the gym on a weeknight sweating profusely. It is at this point that you will need to fall back on your hard motivations, which might be to lose 10 kilograms because the doctor told you that you are heading for a heart attack if you don't, to get you through the rest of the session.

Soft motivations are used in the army as a means of helping soldiers remain motivated and focused on their tasks. Infantry school trainees are allocated into platoons (usually thirty-two soldiers) which are then divided into four sections (eight soldiers). When doing the obstacle course, which is undertaken at least weekly, the section with the fastest time is rewarded by being excluded from the Friday morning barracks inspection. This isn't the reason why the soldiers joined the army in the first place. It does, however, act as a small incentive and gives the soldiers a motivational boost when they are going through the course wearing gas masks (which make breathing difficult) after having just completed a 10-kilometre forced pack march.

Examine and find your own soft motivations. You are likely to have many that have inspired you at some point. Think about what has inspired you in the past and write them down. Look at ways you can apply them to the things you do in your life in a manner that can help motivate you when you are struggling to remain disciplined. Keep in mind, however, drawing upon the same soft motivations repeatedly may dilute their effectiveness, so consider saving these motivations for when your discipline starts to waiver, and use it as a tactical resource rather than an everyday thing.

Hard motivations and soft motivations work together collaboratively. However, as I've stated before, we are all only human. At some point, due to mental and physical fatigue, even hard motivations will start to waiver, and so it's perfectly natural to question why it is you are going through the hardships needed to achieve your goals. At this point, you need to rely on your self-discipline. However, you can use a soft motivation to help reignite that spark to give yourself a much-needed boost to reinvigorate your efforts. Motivators can help you do this.

Motivators

Motivators remind you of your motivations. Motivators assist you to keep disciplined because they serve to remind you of why you are chasing your goals. They are the source of an entire business industry, whether it be motivational books, movies, music, quotes or podcasts. They are usually audio or visual tools that you have in place to help remind you of what your motivations are, so they can be readily available if you need to call upon them. Sometimes memories or particular recollections are the basis of your motivations. Motivators are there to remind you of why you are chasing a goal or doing something to begin with. They are usually very simple things, such as a song on your personal music player, a 'before' or 'progress' picture on the fridge to encourage you to keep up your fitness training or a written quote you have on a Post-it Note in your diary that inspires you to work hard every day. It could be a particular scene from a movie that inspires you to keep focused.

Almost everyone who goes into the gym has a personal music player, which is full of music or audio pieces that they like to train to. It might be music of a high-beat tempo, something to really get the blood going, or it could be something tamer, that helps the person undertaking training relax and focus on their session, clear of any other thoughts. Part of the showmanship of boxing is the fighter's walk to the ring, and it is accompanied by loud music. This serves two purposes: to maintain a party-like atmosphere in the crowd, and to motivate the pugilists as they prepare to enter the ring. It's an adrenalin rush that helps to motivate them just before the fight.

All motivators need context. This means that they will be things that inspire and drive you because there is a reason or justification behind it. It relates directly to your motivations, whether they be hard or soft. I once worked with an American soldier who kept a small postcard-sized mural of the twin towers in his equipment. This was a personal motivator that reminded him why he was thousands of kilometres from home living and working in a warzone. It had significant context behind it. His motivation was a hard motivation, yet despite a deeply held belief in what he was doing, he still kept the small motivator because human nature means that our discipline and motivations will waiver from time-to-time. It can be as basic as having your friend encouraging you when in the gym, motivating you because you have set a goal that you each want to gain 5 kilograms of muscle before summer. It could be a piece of music which reminds you of a movie or story that you found particularly inspirational and is what made you want to set out to achieve a particular task or goal in the first place.

There are plenty of motivators that you can use. People generally don't like to let their family and friends down, and they certainly don't like to feel foolish. One of the most effective motivators you can put in place is telling those close to you that you intend to set out to achieve something. It could be to do a course of study or train for and run a marathon. Telling people about your goals serves two purposes. Firstly, it means that others are now aware of your intentions, so you are far less likely to start working towards something and then throw in the towel when things start to become difficult or you struggle to find motivation and discipline. You will want to avoid telling your family and friends that you gave up, because this will make you feel silly and perhaps even be the subject of good-natured ridicule, which no one really likes. Fear of failure is a powerful motivator. Secondly, it's likely that your family and friends will want to support you in your endeavours, so you can leverage the good feeling that comes from their encouragement and congratulations as you keep achieving milestones and working towards your goals. They can be your moral support when you are feeling unmotivated and struggling to find discipline. If you have to delay your goals or change them for reasons outside your control, then they will know you have tried and will be there to help you refocus.

A 'conquer or die' approach was used by Hernan Cortez, a Spanish conquistador who, in 1519, ordered that all the ships that had been used to sail to Veracruz be scuttled (various historians argue that they were in fact burned). The Spanish were then highly motivated to succeed in their conquering of Aztec tribes, as they knew the stories of the Aztec's brutal slayings of their adversaries.

The army uses many small motivators to help achieve objectives, especially when it comes to training personnel. As mentioned earlier, Australian trainee infantrymen get rewarded for small wins, such as setting a superior time on the obstacle course which is rewarded by not having to participate in a barracks inspection. Additionally, these same soldiers are encouraged to put up paraphernalia in the lines to help keep them motivated. Many of my friends, including myself, had a strong desire to make it into the airborne battalion on completion of training. To help motivate us, we had pictures of paratrooper's scattered all over the walls of the bathrooms to remind each of us of what we were working towards achieving. Our motivations for being in the army and putting ourselves through such strenuous training varied, but a common motivator helped each of us push through and obtain our goal.

When undertaking a long pack march during one of the phases of special forces selection, alone and in the middle of a cold night, a friend of mine recalled how he thought about what was waiting for him back at his parent unit if he failed. Knowing that he would go back to painting rocks and standing in the guard box (all unfortunately a part of the life of conventional soldiers when in barracks) was enough to help motivate him. His motivation to serve in such a unit was already very strong, but something as simple as this thought helped him remain disciplined through the selection course.

Before going 'outside the wire' (the term used for when military units leave the relative safety of their bases and go out into hostile areas) on patrols or direct-action missions, many soldiers like to listen to some sort of inspiring music to help motivate them. In the movie *Blackhawk Down,* there is a scene where soldiers are preparing to go on their mission to capture a warlord. As they prepare to leave the safety of their base and enter what they know will be hostile territory, music by the band Faith No More can be heard blaring from the loudspeakers in the compound. Some people have a quote or a picture above their door that they like to refer to before they begin their day. It will usually have some meaning to the individual, or it may just be some inspiring words that gives them some thought to the activity they are about to undertake.

One of the best motivators that you can have is a 'mantra'. A mantra is a basic statement or slogan that is used to concentrate and to exemplify your motivations. You can use one of the literally thousands of quotes or sayings in existence as a mantra for your own life or your particular goals, or you can create one for yourself. Your mantra might be a picture, such as my American colleague with his postcard mural. A mantra is designed to have a deeper meaning that reminds you of all the reasons you are doing the things you are, and it is also designed to discourage you from giving up easily. It's aim is to make you think positively. Remember, it needs to have context, otherwise it will just be words (or a picture) that will serve no purpose when you are really looking for motivation to help keep you on track towards achieving your goals.

I developed a mantra after I had finished my military career and commenced work in the private sector. When many servicemen and women finish military service, whether by choice or due to circumstances beyond their control, they find themselves lacking direction and losing motivation very quickly. I was no different. I found myself struggling to get out of bed each day, to go to the gym as frequently and with the same purpose that I had when I was in uniform. I struggled to find significance in the new civilian job I had taken on. I still went to the gym and went to work because I had developed a high degree of discipline. However, I was unmotivated. I discussed this with some of my friends, many of whom had been wounded in combat or during training accidents. During these discussions, a lot of what was hazy and unclear to me started to make sense. One afternoon after one such chat, I went home and wrote some words onto a small yellow Post-it Note and placed it on my fridge. It simply said, 'Because you can'.

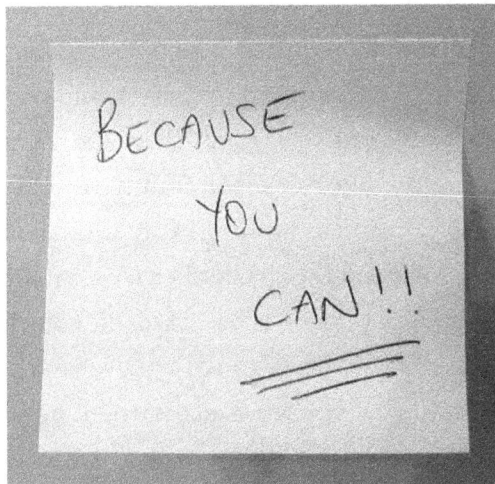

These words, which may mean nothing to others, have significant context for me because it reminded me not to think myself hard done by. I, unlike many other soldiers who served in conflict, was fortunate enough to return home, mostly in one piece. I remembered that I should be grateful for this, and it motivated me to try and take a 'glass-half-full' approach to life once again. I still had things I wanted to set out to achieve, but I had been lacking some motivation. Having this mantra helped me because when I was struggling to get out of bed, I told myself, 'Get up because you can, unlike those who will never wake up again'. When I was in the gym and struggling with a particularly hard legs session, I would tell myself, 'Do the extra set because you can, unlike the soldiers who lost their legs to improvised roadside bombs'. It was designed to remind me of what could easily have been a different life and to cherish the things I can do. This became my motivator for everything.

Look to have something similar for yourself, but ensure it has context. After giving a speech on motivation at a veteran's function one night, I had an ex-sailor approach me and tell me that he keeps his old Centrelink card (the Australian welfare system for unemployed people) in his wallet as a motivator to help deal with the lengthy bus trip into his job each day. It motivated him to do his job well, as he never wanted to have to rely on welfare again (though he repeatedly emphasised he had been grateful that the safety net had been there for him). Seek out motivators that work for you, it may be a single one or several. There are numerous stories and examples on the internet about people who have achieved great things despite great adversity that can be very inspirational. Have them handy and available when you need them, whether it be a memorable quote on the bathroom mirror or a picture of your kids in your wallet motivating you to be the best person you can be and to achieve your goals! Remember, they will be relevant to you, your own experiences or your own perspectives, so don't be dismayed if other people don't understand them as well as you do.

SUMMARY POINTS

- Motivation is a valuable tool that is the reason you do the things you do.
- Hard motivations are what drive you.
- Soft motivations are what inspire you.
- Motivators are the triggers that make you think about your motivations, and you use them when working towards your goals.

WRITE DOWN YOUR HARD AND SOFT MOTIVATIONS

WHAT ARE YOUR OWN MOTIVATORS?

Discipline *and* Motivation: The Key to Success

"When that first bullet goes over your head,
politics and all that shit goes right out the window"

'Hoot'—from the movie *Blackhawk Down*

As stated earlier, discipline and motivation leverage off each other and should be viewed as complementary tools. Discipline is the tool that allows you to maintain the effort towards achieving your goals, while motivation is what drives and inspires you. For example, motivation can help you renew your work efforts when you are struggling to keep disciplined, and discipline is how you keep on task when motivation is wavering. They complement each other, so it is important to determine what your own motivations are and to instil discipline in yourself.

Having a plan is a necessary part to setting out what it is you want to achieve and do, and this will likely stem from your motivations. However, without discipline, a plan is just a statement of intent. Discipline is that bridge between desire and achievement, because motivation will trigger an intent to achieve a goal and help to get you going initially. Discipline will allow you to keep focused towards the completion of that goal and is the extension of your motivation, regardless of how strong or weak that motivation is. Ensuring that your motivations and discipline are synchronised will allow you to achieve anything you set out to do!

A modern example of the success of the combination of discipline and motivation was the efforts by US special operations forces during the Battle of Mogadishu, which occurred in Somalia on 03 October 1993. When an ostensibly straight-forward operation went very wrong after a supporting Blackhawk helicopter was shot down near the target area, US soldiers engaged in a two-day battle to fight their way through heavily hostile urban terrain and back to the safety of their operating base on the outskirts of the city. This required an inordinate combination of discipline and motivation in order to simply survive. Their discipline stemmed from their many years of training. This allowed them to fight their way tactically, block-by-block, to ensure the injured soldiers were looked after and cared for during the battle. The easier option could simply have been to send more helicopters in to conduct a rescue, but it was assessed that this would have led to more casualties.

Their motivations were hard. The raid was designed to capture a warlord who had been responsible for the death of two dozen United Nations peacekeepers months earlier. Their initial motivations were to assist the local population by helping to bring peace to the country and to bring the warlord to justice. However, once the operation went sour, their motivations were solely to get out of the situation alive and to ensure as many of their injured colleagues as possible would live. This was a strong motivation, but the almost herculean effort of getting out of the kill zone and back to safety could not have been achieved if they hadn't maintained their very high level of discipline.

Similar concepts of changing motivation are found in the sporting world. In team sports, the members of a team losing in the final phases of a match, with no chance of winning, will need to find different motivations from when the match commenced. Some players will struggle to find motivation in a losing cause and may seek to preserve themselves for the next match and give half-hearted efforts, often to the disdain of the viewing public, their teammates and the coaching staff. Others will use it as an opportunity to excel individually, perhaps endearing them to a coach who will now be seeking to make player changes for the following match. The most respected, and usually best, players are the ones who keep disciplined and motivated, even if all is lost, as they are driven by a desire to know that they gave their best performance. This will help motivate the younger and more inexperienced players to never give up and will help instil discipline in those players so that when they are involved in a match that is more competitive, they have the experience of giving their best efforts right until the end of the match.

Whenever working towards something, always ask yourself, 'Why am I doing it?' If you don't have a good reason, you are likely wasting your time. What you then need to ask yourself is, 'What is motivating me to do this?' If it is a hard motivation, you shouldn't have too much difficulty applying yourself towards getting the task done. However, even with a hard motivation pushing you to pursue a goal, especially one that may take a long time to complete, such as a twelve-week weight loss program, motivation can sometimes waiver. This is where you need to call upon your discipline, and consider supporting it with soft motivations. For example, playing some inspirational music on your headphones that helps get you into that cardio session at the end of a hard day when you would rather just sit on the couch. Even when highly motivated, you still need to remain disciplined to ensure that you are doing everything you need to do to achieve your goals. Complacency breeds mediocrity. It can be the cause of a military mission that is succeeding suddenly failing.

The ability of the terrorist organisation Daesh to capture large swaths of territory in northern and western Iraq in early 2014 was the result of the Iraqi Army, a superior-equipped and trained force, becoming complacent and believing there was no threat posed by that organisation. When Daesh began to attack the Iraqi forces, and started claiming victories, the Iraqi soldiers began to panic, and discipline became virtually non-existent. Compounding this was the fact that many of these soldiers were not motivated to fight, as Daesh was taking over areas that weren't part of their tribal homelands (Iraq, by its very nature, is made up of many long-standing tribes, all based in various parts of the country, some which overlap the borders created by European powers). It wasn't until Daesh was within a day's drive of Baghdad that the coalition forces returned to Iraq at the request of the Iraqi government, which was now motivated by the desire to defend their capital.

Knowing your own motivations and discipline requirements (and limitations) will let you incorporate them together when working towards your goals. This will ensure that complacency doesn't creep into your own approach when working towards them. It will allow you to keep working hard even when things get tough. It's all mental acumen. Your mind will give in well before your body does. You can be highly driven to lose weight and do better in the gym, and your motivation can stem from a deep desire to feel better about yourself. However, if over time you start becoming lazy in the gym, thinking that simply turning up and going through the motions on the treadmill is enough to achieve your goals, then you will fail. Mentally, you may simply not feel like doing what is required. Or you may start to believe that doing the bare minimum will still be enough to achieve your goals. This is simply human nature. This wavering of motivation, and consequently effort, can happen among all your focuses, whether it be fitness goals, workplace goals or personal goals. Even if you have planned adequately, there will still be times where you need to push through the 'boring parts'. Knowing that you have discipline, learnt by having practiced it and having instilled it into yourself, will help you avoid becoming apathetic or producing half-efforts when striving towards the things you want to achieve.

This unconscious depreciation of motivation is where your discipline needs to step in. This will let you complete the full five rounds of the high-intensity circuit in the gym instead of stopping after just four. It will allow you to complete the work project to a consistently high standard instead of rushing it at the end. Discipline will ensure that an infantry platoon doesn't stop their tactical movement when they're on patrol simply because they're almost back inside the relative safety of their operating base.

You should visualise what achieving your goal looks like. Imagine yourself in the position that you are trying to be in, so that it feels more real, which will make all the hardships more bearable and give you a picture, if only in your head, of what you are working towards. The combination of discipline and motivation is the key to achieving the things you never thought possible. They can feed off each other so when chasing your goals becomes difficult, which it will be at times, you will have that voice in your head telling you to stay disciplined enough to keep going. That voice will remind you that you have reason and purpose for pursuing them, and you have the ability to maintain the effort required even when it begins to look too hard.

SUMMARY POINTS

- **Discipline and motivation complement each other.**
- **Discipline is the key to achieving your goals!**

Conclusion

Before initiating your own C2 Framework, I highly recommend doing a review on where you are in life right now. Write it all down. From here you can articulate your focuses, assign goals to them, determine a battle rhythm and better understand your own motivations and discipline. Ask some basic questions of yourself. What are the things in your life you devote your time to? What do you want to achieve? What do you want to become better at? How are you spending your time? What can you remove from your life that wastes your time? What are your motivations? Are you disciplined? Could, with a little effort, you become disciplined? (The answer to this last one is always a resounding 'yes'. The question is, how bad do you want it?)

The C2 Framework is a simple methodology, so anyone can use it. When looking to instil more discipline in your own life, don't feel threatened or scared of it, embrace it instead. You already have discipline; it's taught to you as a child by things like looking both ways before crossing the road. It's the level of discipline, and what you can achieve with it, that you need to look at. When looking at your motivations, you are simply examining those things internal to you that already exist. The next stage is harnessing them so that you can set out to achieve goals and be more efficient with how you plan and use your time to pursue them. If your motivations aren't particularly strong, do some research and seek to instil a mantra of your own.

Many soldiers are experts at discipline and motivation as a direct consequence of the events and situations they have experienced in line with their work. After months (and in some cases years) of operational service in some of the most hostile environments on Earth, everything else after that seems much easier. For them, being disciplined and leveraging motivation is second nature. However, you don't have to go through these same experiences to learn these lessons for yourself. Draw upon your own experiences or upon those of people close to you. Look to the examples of people who inspire you through their achievements or life stories. It's up to you to decide what works best and to establish the framework in a way that will allow you to achieve the things you want. Even if this is the first time you are setting goals and using a time management plan, simply doing so is the first step in the right direction.

Everyone has different levels of ambition. Some people want to achieve many things and to a high level, while others are more content to lead simple lives that are focused on smaller achievements. The latter is no less significant than the former, as our individual goals are important to us and us alone. It's always a good idea to try and gain the support of family and friends to pursue your goals, but this might not always be forthcoming. This is why motivation and discipline are critical to chasing your goals. There is no greater satisfaction than setting out to achieve something, dealing with the joys and hardships of that pursuit, and then reaching that accomplishment, whether it be big or small. You are ultimately responsible for taking the direction of your life into your own hands. This book is simply designed to act as a guide to help you have structure and direction in your life, to use your time wisely, and to use your motivations and discipline to achieve your goals. It's now up to you to do it!

About the Author

Josh Francis is a former Australian Army soldier, having served in Afghanistan, Iraq and East Timor. A qualified high school teacher, he developed Red Diamond consultancy as a means of providing motivational and leadership advice to individuals and groups.

For more information visit

www.red-diamond.com.au

or contact

director@red-diamond.com.au